MARINE CORPS AVIATION:

THE EARLY YEARS

1912-1940

by

Lieutenant Colonel Edward C. Johnson, USMC

Edited by

Graham A. Cosmas

HISTORY AND MUSEUMS DIVISION

HEADQUARTERS, U. S. MARINE CORPS

WASHINGTON, D. C.

1977

FOREWORD

This brief history of Marine aviation from 1912 to 1940 describes the efforts of Marines to secure their own air arm and recounts the early development of the Marine air-ground team. The story is drawn from official reports, documents, and personal correspondence, as well as from published historical works. It also draws heavily upon the transcribed reminiscences of notable Marine aviators collected and preserved by the Oral History Section of the History and Museums Division.

Lieutenant Colonel (now Colonel) Edward C. Johnson, USMC, did the initial research for this history and wrote the first draft. Colonel Johnson received his Bachelor of Arts degree from the University of Wisconsin and is himself an experienced fighter pilot, squadron and group commander. He commanded VMFA-251 in 1969 when the squadron received the Robert M. Hansen Award for outstanding performance. Colonel Johnson came to the History and Museums Division in June 1970 from Vietnam, where he served on the staff of Marine Aircraft Group 13.

Additional research and editing of the manuscript were done by Carolyn A. Tyson and Captain Steven M. Silver, USMCR. Dr. Graham A. Cosmas extensively revised the manuscript and incorporated in it much primary source material not available when Colonel Johnson prepared the initial draft. Dr. Cosmas, who received his Bachelor of Arts degree from Oberlin College and his doctorate from Colonel Johnson's *alma mater*, the University of Wisconsin, completed the editing of the manuscript and prepared it for publication.

The History and Museums Division welcomes any comments on the narrative and additional information or illustrations which might enhance a future edition.

E. H. SIMMONS
Brigadier General, U. S. Marine Corps (Ret.)
Director of Marine Corps History and Museums

Reviewed and approved:
1 August 1977

PREFACE

From 1912, when First Lieutenant Alfred A. Cunningham became the first Marine to fly, through 1940, a handful of dedicated Marines worked to keep their Corps abreast of the progress of military aviation and to create an air arm specifically dedicated to supporting Marines in their amphibious mission.

From a few daring men and a handful of primitive aircraft in 1912, Marine aviation grew into a force which met the test of combat in World War I. During the 1920s and 1930s, Marine aviators gradually developed a permanent organization and acquired aircraft of increasing reliability and improving performance. In small wars and expeditions in Haiti, the Dominican Republic, Nicaragua, and China, Marine fliers devised new techniques for supporting Marine infantry in combat, and they demonstrated the value of aviation in reconnaissance and in the movement of men and supplies over rough and usually roadless terrain.

With the creation of the Fleet Marine Force in 1933, Marine aviation received formal recognition as an element of the amphibious air-ground team, and in the fleet landing exercises of the late 1930s began developing the doctrines and tactics which would make close air support a reality in World War II. The traditions of excellence and versatility established by these early Marine fliers lived on in the skies of Korea and Vietnam and remain vital today.

This study of the formative years of Marine aviation is based on official reports and documents in the archives and holdings of the History and Museums Division and on personal memoirs and correspondence, as well as published historical works. It draws heavily on the writings of such pioneers of Marine aviation history as Robert L. Sherrod and Major Edna Loftus Smith, USMCR, and has benefited significantly from the efforts of such organizations as the First Marine Aviation Force Association and the Marine Corps Aviation Association to preserve the memory and record of early Marine aviation.

Especially valuable in recalling this era to life were the oral reminiscences of distinguished retired Marine aviators transcribed and preserved by the Oral History Section of the History and Museums Division. Among others, the recollections of General Christian F. Schilt, Lieutenant Generals Karl S. Day and Francis P. Mulcahy, Major Generals Ford O. Rogers, Lawson H. M. Sanderson, and Louis E. Woods, and Brigadier Generals Edward C. Dyer and Daniel W. Torrey enriched the narrative.

The division owes a special debt of gratitude to those persons who furnished assistance, comment, and criticism on the initial draft of the history. Among them, Master Sergeant Roger M. Emmons, USMC (Ret.), Historian, Marine Corps Aviation Association, commented on the manuscript and furnished many valuable documents on World War I Marine aviation; we have borrowed much from his earlier writings on the period. Mr. Lee M. Pearson, Historian, Naval Air Systems Command, gave us extensively of his time and of his knowledge of early naval aviation. General Vernon E. Megee, USMC (Ret.), provided especially useful factual comments, and Master Sergeant Walter F. Gemeinhardt, USMC (Ret.), member of the staff of the Marine Corps Museum at Quantico, gave us the benefit of his detailed knowledge of early aircraft and the men who flew them.

Thanks are due also to Mr. Goodyear K. Walker of Sacramento, California, for providing the Kirkham photograph albums, selections from which have enriched the illustrations of our history, and to Colonel Houston Stiff, USMC (Ret.) of the Treasure Island Navy/Marine Corps Museum at San Francisco for bringing these albums to the attention of the Director of History and Museums.

This history could not have been written without the generous assistance of many members of the History and Museums Division. The writer and editors owe particular gratitude to Mr. Ralph W. Donnelly and Mr. Charles A. Wood for their aid in locating records and personal papers and to Mr. Benis M. Frank, head of the Oral History Section, for his guidance to these valuable sources. Mr.

Rowland P. Gill and Mr. Jack B. Hilliard searched out photographs to illustrate the text. The manuscript was prepared under the editorial supervision of Mr. Henry I. Shaw, Jr., Chief Historian, History and Museums Division. The manuscript was prepared for publication by Mr. Paul D. Johnston. Unless otherwise indicated, all photographs are from official Marine Corps/DOD holdings.

GRAHAM A. COSMAS

E. C. JOHNSON

TABLE OF CONTENTS

CHAPTER I.

THE BEGINNINGS, 1912–1917

Naval Aviators in a Different Uniform

Until the United States entered World War I, Marine Corps aviation had no permanent organization separate from naval aviation, and its history is interwoven with that of the rudimentary naval air arm. The first recorded Navy Department expression of interest in heavier-than-air flying machines dates back to 1898, when Assistant Secretary of the Navy Theodore Roosevelt, impressed by reports of the experiments of Professor Samuel P. Langley, tried to promote consideration of the military possibilities of aeronautics. He met a sharp rebuff from the Navy Department bureaus. For the next 12 years, while the Wrights flew at Kitty Hawk and aviation activity slowly increased in America and Europe, the Navy cautiously observed developments.

In 1910, a year after the Army bought its first plane from the Wright Brothers, the Navy Department assigned Captain Washington Irving Chambers, a veteran sea officer long interested in aircraft, to answer correspondence concerning aviation. Chambers had neither authority nor a staff, but he set out to awaken the Navy's interest in flight and to promote aeronautical research. He obtained important allies within the Navy Department, including the venerable Admiral George Dewey, and he developed a close working relationship with the aircraft builder and inventor Glenn Curtiss. On 14 November 1910, as a result of Chambers' and Curtiss' joint efforts, Curtiss' test pilot Eugene Ely made the first recorded takeoff from a ship's deck, flying from a platform erected on board the U.S.S. *Birmingham*. Two weeks later, Curtiss offered to teach a naval officer to fly at no cost to the government. The Navy Department accepted, and Lieutenant Theodore G. Ellyson, USN, reported for instruction to Curtiss' flying school and experimental station at San Diego, California. There he soon qualified as the Navy's first officer pilot.

The following year, naval aviation acquired more personnel and its first aircraft. With a Congressional appropriation of $25,000, the Navy Department in 1911 purchased three planes—two from Curtiss and one from the Wright Brothers. The manufacturers trained an officer pilot and an enlisted mechanic for each aircraft, including Lieutenants John Rogers and John H. Towers, and they in turn began training others. Chambers secured establishment of a primitive aviation camp (a field, a few buildings, and a beach for launching sea planes) near the Navy Engineering Experiment Station at Annapolis, just across the Severn River from the Naval Academy. Ellyson, Rogers, Towers, and their enlisted mechanics began flight operations there in the fall of 1911. They combined training with experimental work. To avoid winter weather, they moved to San Diego in December and wrecked all three of their airplanes while flying from the Curtiss field. The following spring they returned to Annapolis. At a new site (the previous year the camp had been in the line of fire of the Naval Academy's rifle range), they began rebuilding their aircraft and prepared to train the new aviators who soon came to join them.

Among the prospective aviators who reported to the Annapolis camp in 1912 were two Marine officers. Their presence reflected a slowly growing Marine Corps interest in aviation. This interest was closely related to the emerging new mission of the Corps in the steel battleship Navy: occupation and defense of advance bases for the fleet. Since 1900, the Navy and Marine Corps had been trying to organize an Advance Base Force of Marine infantry and artillery. Shortages of funds and manpower and lack of agreement on details of its organization had hampered the actual formation of the force, but by the end of 1911 an Advance Base School was

Lieutenant Alfred A. Cunningham cranks up "Noisy Nan" for a test flight at Philadelphia in 1911. The plane's inventor is at the controls. (Marine Corps Photo 514941).

in operation at the Philadelphia Navy Yard. During 1912, the Marine Corps staff in Washington concluded that, in the words of Major General Commandant William P. Biddle, "great benefit to an advanced base force . . . might result from trained aviators."[1] Accordingly, the Marine Corps ordered two officers then assigned to the Advance Base School, First Lieutenants Alfred A. Cunningham and Bernard L. Smith, to Annapolis for flight instruction and aviation duty. With this routine order began the epic of Marine Corps aviation.

For Alfred A. Cunningham, the first of the two Marines to arrive at Annapolis, the assignment represented the fulfillment of a dream long pursued. Born in Atlanta, Georgia, in 1882, Cunningham enlisted in a volunteer infantry regiment during the Spanish-American War. He was mustered out of the Army after a tour of occupation duty in Cuba, returned to Atlanta, and spent the next 10 years selling real estate. During these years in Atlanta aviation caught his imagination, probably in 1903 when he made his first flight of any kind—a balloon ascent. In 1909, he resumed his military career by accepting appointment to the rank of second lieutenant in the Marine Corps. After two years of routine duty with battleship detachments and at various shore stations, he was promoted to first lieutenant in September 1911 and in November of that year he reported to the Marine Barracks at Philadelphia for duty and instruction at the Advance Base School.[2]

Cunningham had retained his interest in aero-

nautics, and at Philadelphia he found an active aviation movement among civilians and off-duty military personnel. Cunningham soon joined the unofficial experimenters. With his own money, he rented an airplane from its builder (who needed the $25 per month to buy food) and persuaded the commanding officer of the Navy Yard to let him use an open field on the base for test flights. Even Cunningham's enthusiasm, however, could not overcome the aerodynamic deficiencies of "Noisy Nan," as he called his rented aircraft. The young, aspiring aviator described his days of frustration: "I called her everything in God's name to go up. I pleaded with her. I caressed her, I prayed to her, and I cursed that flighty old maid to lift up her skirts and hike, but she never would."[3*]

Besides struggling with "Noisy Nan," Cunningham joined the Aero Club of Philadelphia, the city's principal organization of aviation enthusiasts, and he launched a campaign to interest the Marine Corps in establishing its own aviation force. Evidently making good use of his experience selling real estate, he sold the members of the Aero Club, many of whom were wealthy, influential Philadelphia socialites, on the idea that their city should have a Marine air

* "Noisy Nan" probably was underpowered for its weight and might well have proved highly unstable had Cunningham managed to get it into the air. It was an excellent preliminary trainer, however, giving Cunningham the "feel" of flying in its takeoff runs and occasional brief hops off the ground. (Gemeinhardt Comments)

base. The Aero Club members, through their political friends in Washington, D.C., brought pressure to bear on a number of officials, among them Major General Commandant Biddle, himself a member of a prominent Philadelphia family.[4]

What effect, if any, Cunningham's unmilitary methods of advocacy had on the decision to detail Marine officers for pilot training is a matter of conjecture; but his activities at least appear to have assured him first place on the list of potential aviators. On 16 May 1912, Cunningham received orders detaching him from the Marine Barracks at Philadelphia and instructing him to report on 22 May to the U.S. Naval Academy at Annapolis (which meant the nearby aviation camp) "for duty in connection with aviation." [5]

Cunningham reported at Annapolis on the specified date, only to be ordered away almost at once on expeditionary duty. When he returned in July, no aircraft were available for him to fly. Eager to begin flight training at once, he obtained orders to the Burgess Company and Curtiss factory at Marblehead, Massachusetts, which built the Navy's Wright aircraft and which had airplanes and civilian instructors.[6] There, after two hours and 40 minutes of instruction, Cunningham soloed on 20 August 1912. He later explained the brevity of his training and described his solo flight:

> There being so few civilian flyers, the factory had to pay them a huge salary to teach us, and they were anxious to make it short and snappy . . . I had only attempted to make two landings in rough weather when one calm day they decided to risk the plane rather than continue to pay any instructors large salaries. I was asked if I was willing to try it alone, and said I was. I took off safely and felt confident in

the air until I thought of landing and wondered what would happen when I tried to do it alone. Every time I decided to land I would think of some good excuse to make another circle of the bay. The gas tank was mounted between the wings in plain view, and a small stick attached to a float protruded from the top of it for a gasoline gage. As the gas was used, this stick gradually disappeared within the tank . . . As this stick got shorter and shorter, I became more and more perturbed at having to land with little idea of how to do it. Just as the end of the gasoline gage stick was disappearing, I got up my nerve and made a good landing, how I don't know . . . This was my first solo.[7]

Cunningham eventually was designated Naval Aviator No. 5 with the effective date of his designation arbitrarily set as 17 September 1915.* Both the date of his solo and the date thus fixed for his formal recognition as a naval flier have their advocates as "birthdays" of Marine Corps aviation, but the date he reported to the Aviation Camp at Annapolis, 22 May 1912, is the officially accepted birthday.

---

* Because the Navy was slow to establish official performance standards for aviation, precedence numbers and dates of designation of the first aviators, including Cunningham, are confusing and arbitrary. The Navy set its first official aviation performance standards in April 1913. Almost two years later, the Secretary of the Navy sent letters to fliers designating them as Navy Air Pilots and setting precedence dates. This list omitted the names of Ellyson, Rodgers, and Cunningham—Naval Aviators 1 and 2 and Marine Corps Aviator 1—because they were not on aviation duty at the time the letters were issued. In 1918, when golden wings were distributed as the aviators' official badges, this omission was rectified. All naval aviators received new precedence numbers including the earliest fliers, whose numbers were based on the order in which they reported for flight training. Cunningham thus became Naval Aviator Number 5, with his qualification date arbitrarily stated as 5 March 1913, and his date of designation as a naval aviator as 17 September 1915. (Pearson Comments: Caidin, *Golden Wings*, Appendix A).

Commonly called the "Bat Boat," the Wright B-1 seaplane was the third aircraft owned by the U.S. Navy and in 1912 was used to train Marine aviators. (Marine Corps Photo 514903).

On 18 September 1912, First Lieutenant Bernard L. ("Banney") Smith, the second Marine assigned to aviation training, arrived at Annapolis. Born in Richmond, Virginia, in 1886, Smith had entered the Marine Corps in the same year as Cunningham and was only a few days junior to Cunningham in rank based on the dates of their promotions to first lieutenant. By the time Smith reported for training, the Navy's three aircraft had been repaired. Towers, therefore, taught Smith to fly in one of the Curtiss machines, the A–2. Apt and enthusiastic, Smith soon soloed and flew frequently with Towers and Ellyson. When naval aviator designations were sorted out, he went on the list as Number 6.[8]

In September, Lieutenant Ellyson, now officer in charge, reorganized the aviation camp, assigning particular officers to each of the four aircraft. Cunningham received the B–1, the older of the two Wright machines, while Smith took charge of a Curtiss, the A–1. Naval officers flew the other two planes. Cunningham, with Sergeant James Maguire, the first enlisted Marine assigned to aviation duty, became known informally as the "Marine Camp," while Smith worked with Lieutenant Towers in what was called the "Curtiss Camp." *

Throughout early 1912 and 1913, the naval aviators continued to conduct training flights and tests of tactics and aircraft capabilities. They experimented during this period with detection of underwater objects from the air and with air-ground radio communication. In January 1913, the aviation camp for the first time joined the fleet in its annual maneuvers off Guantanamo, Cuba. The aviators here proved that they could locate submerged submarines and that they could spot enemy surface vessels without themselves being sighted. They dropped missiles from the air and took photographs. In an effort to increase interest in aviation, they carried over 150 Navy and Marine officers on indoctrination flights. One of these officers, who flew with Lieutenant Towers, was a future Commandant of the Marine Corps, Lieutenant Colonel John A. Lejeune. He spent 14 minutes aloft.[9]

Cunningham and Smith both participated in these activities, but Cunningham was hampered by the inadequacies of his plane. The B–1, the first Wright Brothers aircraft purchased by the Navy, was powered by a single engine which drove twin propellers by long chains connected to sprocket wheels. It had been wrecked and rebuilt several times before Cunningham took it over, and Cunningham rebuilt it again. The performance of the aged machine steadily deteriorated, as Cunningham reported to Captain Chambers:

> My machine, as I told you and Mr. Towers probably told you, is not in my opinion fit for use. I built it from parts of the Burgess F and Wright B, which are not exactly alike and nothing fitted. I had to cut off and patch up parts and bore additional holes in beams in order to make them fit. The engine bed, made by Burgess, was not exactly square with the front beam, so the engine had to be mounted a little out of true (with reference to the engine bed). I have made over 200 flights in this machine and recently, in spite of unusual care of myself and men, something seems to vibrate loose or off a majority of the flights made. One of the propeller shafts is the same one used with the Cyro motor in the old machine. It is the only left-hand shaft here. While the engine runs smoothly, it does not deliver nearly as much power as when it was newer, and even then, it did not have enough power to fly safely in any but smooth weather. It is impossible to climb over a few hundred feet with a passenger. The whole machine has just about served its usefulness and I would like very much to have a new machine of the single propeller type. Lt. Arnold, of the Army, after seeing the machine run and examining it, said that none of the Army fliers would go up in it. Will you kindly let me know what the prospects are for my getting a new machine.[10] **

In spite of the B–1's faults, Cunningham managed to make almost 400 flights in it between October 1912 and July 1913. On a couple of occasions, his craft stayed airborne long enough to cover about 80 miles and it reached a maximum altitude of around 800 feet. The more frustrating days were typified by the terse entry on the page of Cunningham's flight log recording flights number 371 through 383: "Engine stopped in air on nearly all these flights." [11]

In August 1913, Cunningham requested and received detachment from flight duty. He stated concisely his reason for requesting this transfer: "My fiance will not consent to marry me unless I

---

* This might be considered the beginning of a divergence in the careers of the first two Marine aviators, a divergence which steadily became more apparent. Cunningham from the start was not only an aviator but emphatically a *Marine* aviator, always promoting a distinctive Marine air entity. Smith, on the other hand, identified himself more generally with naval aviation and contributed much to its development.

** The "Lt Arnold" referred to was the future General Henry H. ("Hap") Arnold, commander of the U.S. Army Air Forces in World War II. One of the early Army aviation stations was located at College Park, Md., so that informal contact between the Army fliers and their Navy colleagues at nearby Annapolis was frequent. (Van Deurs, *Wings for the Fleet*, p. 51.)

give up flying." [12] * Assigned to ground duty at the Washington Navy Yard, Cunningham continued to advocate Marine aviation and soon would make some of his most valuable contributions to it.

After Cunningham's departure, Lieutenant Smith continued flying with the Navy aviators. He was joined in November by the third Marine to be assigned to aviation, Second Lieutenant William M. McIlvain. McIlvain soloed the following month and became Naval Aviator Number 12.

At the end of 1913, the Navy's air arm consisted of 8 aircraft with 13 qualified officer pilots. Of this number two (not counting the grounded Cunningham) were Marines, and seven more enlisted Marines were in training as mechanics. In October, the Major General Commandant recommended that the Marine aviation personnel "with the necessary equipment" be moved to the Philadelphia Navy Yard for duty with the advance base regiment then being assembled there. [13] This recommendation soon would be followed by the first tactical deployment of Marine aviators with Marine ground forces.

Slow Steps Forward, 1913–1917

During 1913, while the pilots at Annapolis flew, repaired their planes, and flew again, a series of Congressional and Navy Department actions placed naval aviation, and Marine aviation as a part of it, on a more solid organizational foundation. In the Naval Appropriations Act for fiscal year 1914, passed on 4 March 1913, Congress authorized an increase in pay of 35 percent for officers actually flying heavier-than-air machines. The same legislation limited the total number of Navy and Marine aviators to 30 and provided that none could hold rank above that of lieutenant commander or major. In spite of these limitations, the act constituted the first formal recognition of the air service as a separate specialty for Navy and Marine personnel. On 31 August, the General Board of the

Navy,** the service's planning agency, after a major study of U.S. and foreign aeronautics, called for the creation in the Navy of "an efficient . . . air service" directed in the Navy Department by an officer with full authority on questions of personnel and procurement and with at least captain's rank. The General Board urged the Navy Department to ask Congress for funds for bases, aircraft, and training schools.

In October, Secretary of the Navy Josephus Daniels appointed a board of officers headed by Captain Chambers to prepare detailed plans for the organization of a "Naval Aeronautic Service." Lieutenant Cunningham, temporarily detached from the Washington Navy Yard, represented the Marine Corps on the seven-man panel which included Navy aviators, sea officers, and representatives from the Bureaus of Navigation, Steam Engineering, and Ordnance. After 12 days of deliberations, the board issued a report calling for a force of 50 heavier-than-air craft to be attached to the fleet, with one plane on board each fighting vessel, and special auxiliary ships to carry fuel, spare parts, and extra aircraft. The board also advocated the establishment of a naval air training and experimental station at Pensacola, Florida. Following the General Board's lead, the Chambers Board urged creation of an Office of Naval Aeronautics under the Secretary of the Navy to unify the aviation-related functions then scattered among the bureaus.

Of special interest to Marine aviation were two points in the report. The Chambers Board recommended creation of a separate force of six aircraft "to establish an advanced base ashore," and it suggested that a Marine officer be a member of the staff of the proposed Director of Naval Aviation. Beyond this, the board did not address the organization of Marine Corps aviation or attempt to define its position within naval aviation. [14]

Efforts soon began to implement the Chambers Board's proposals. On 17 December 1913,

---

* There is a common belief that Navy regulations in this period prevented married men from flying, but no such policy apparently existed. Ellyson married in November 1912 but continued flying. Lieutenant John Rodgers actually flew the B-1 with Mrs. Rodgers as a passenger, and other married officers were ordered to aviation duty. Perhaps the future Mrs. Cunningham either flew in the B-1 like Mrs. Rodgers or simply saw the craft. (Pearson comments)

** The General Board was an advisory panel established in 1900 by Secretary of the Navy John D. Long to advise him on plans, policies, and procedures proposed by the bureau chiefs. There was at that time no Chief of Naval Operations, and the bureau chiefs reported directly to the Secretary of the Navy. Although the General Board had only advisory powers, the prestige of its members—who included Admiral of the Navy George Dewey, the President of the Naval War College, the Chief of the Bureau of Navigation, and the Chief of the Office of Naval Intelligence—gave the board considerable influence.

1st lieutenant Bernard L. (Banney) Smith, the second Marine to qualify as an aviator. (Marine Corps Photo 516375).

Captain Mark L. Bristol assumed the post of officer in charge of aviation, replacing Chambers, who retired from the service but remained in the Navy Department as an advisor on aeronautics. By November 1914, an Office of Aeronautics had come into existence within the Division of Operations, with Bristol as its director. Meanwhile, in January 1914, the naval aviation camp moved from Annapolis to an abandoned navy yard at Pensacola. There the aviators and ground crewmen began cleaning up the wreckage left by years of neglect and hurricanes and setting up the hangers and sea plane ramps of what would become the Pensacola Naval Air Station. Late in 1914, as though inspired by the general flurry of activity, Congress included in the annual naval appropriations act $1,000,000 for aviation, to be spent under the direct supervision of the Secretary of the Navy rather than spread among the bureaus.

Early in this year of advance for naval aviation, Marine flyers for the first time briefly attained their own organization and operated with Marine ground units. On 3 January 1914, as the rest of the Annapolis camp prepared to move to Pensacola, a "Marine Section of the Naval Flying School," consisting of Lieutenants Smith and McIlvain with 10 enlisted mechanics, and equipped with a flying boat and an amphibian drawn from the aircraft at Annapolis, embarked at Philadelphia on the transport USS *Hancock*. They sailed for Culebra, Puerto Rico, to join the newly created Advance Base Brigade in the annual Atlantic Fleet exercises.

In the development of the Marine Corps, the Culebra exercise of January-February 1914 had a crucial place. It provided the first test of the Marines' ability to occupy and fortify an advance base and hold it against hostile attack. Landing men, equipment, and heavy guns on Culebra Island, the Marines of the Advance Base Brigade quickly set up their defenses. They withstood simulated bombardments by the fleet and repelled a night amphibious assault. At the end of the exercise, the umpires declared the Marine defenders victorious.

The Aviation Section operated with the brigade throughout the exercise. The aviators set up a temporary seaplane base on Culebra on land blasted clear of trees and mangrove roots. Using their C–3 flying boat because the Owl amphibian* proved too light to carry two men,

---

* The C–3 was a Curtiss "F" Boat. Like other naval aircraft of the time, it was a "pusher," with the propeller mounted on the rear end of the engine and with the pilot

Smith and McIlvain flew scouting and reconnaissance missions. On 22 January, during the fleet's bombardment of Culebra, the Marine flying boat twice circled over the battleships at 5,000 feet altitude, "entirely out of range of small arm fire and (the) high angle of fire making ships' guns ineffective." Lieutenant Smith declared that this feat "shows the possibility of aeroplanes for defense using bombs of high explosive." Almost every day, the aviators took officers of the Advance Base Brigade on flights over Culebra and its defenses "to show the ease and speed of aerial reconnaissance and range of vision open to the eyes of the aerial scout." The Aviation Section left for the United States on the *Hancock* on 5 February. By that time, Smith and McIlvain had made a total of 52 flights, during which they had spent 19 hours and 48 minutes actually in the air.

and passenger sitting side by side in the open air in front of the wings. Both occupants had to lean in the desired direction when they wished to bank and turn the aircraft. The E–1 was an early amphibian, its Owl designation meaning "Over Water or Land." (Gemeinhardt Comments).

In March 1914, on the basis of his Culebra experience, Lieutenant Smith recommended that the Marine air unit for advance base work be composed of five aviators and about 20 enlisted mechanics and ground crewmen. It should be equipped with two flying boats, an amphibian, and a fast single- or two-seater land plane. Smith stressed the need to equip the flying boats, intended primarily for scouting and reconnaissance, with radios, and he urged that canvas shelters for the aircraft and other easily movable ground equipment be provided. Finally, he suggested that the Marine troop transport then under construction be equipped to carry and launch at least one aircraft.[15]

Smith's recommendations dimly foreshadowed later elements of the Marine air-ground team, but immediate reality fell far short even of his modest vision. The Marine air unit ceased to exist at the end of the Culebra exercise and merged once again into the main body of naval aviation at Pensacola. Aircraft and pilots from Pensacola, including "Banney" Smith, participated in the operations at Tampico and Vera Cruz during the Mexican intervention of April

A Curtis C–3 in launch position on the catapult of the USS North Carolina. *The aircraft was assigned to the Marines for maneuvers off Pensacola, Fla., in July 1916. (National Archives Photo 80-G-426924).*

1914, but no separate Marine air unit was created. Smith, stationed with the fleet at Tampico, had no chance to fly in support of the Marine brigade at Vera Cruz.[16]

The outbreak of World War I in the summer of 1914 had little impact on the small band of Marine aviators beyond the temporary detachment from them of "Banney" Smith. The Secretary of the Navy sent Smith to the U.S. Embassy in Paris, where he spent the next two years following and reporting on the explosively rapid wartime development of European aviation. During this assignment, Smith visited and occasionally flew in combat with French air units, and he made a secret trip to Switzerland to obtain aviation intelligence.[17]

During 1915 and 1916, while the war stimulated the growth of European aviation, the advance of American naval aviation faltered. Changes in Navy Department organization in Washington during these years all but abolished the post of Director of Aviation, leaving the air program without a central coordinator or authoritative spokesman. High-ranking officers in the bureaus continued to doubt the military value of aviation and hence failed to press aggressively for its development. At times, they refused to spend money appropriated for aircraft supplies and delayed or prevented the carrying out of legislation.

The continued practical limitations of the available aircraft did much to justify this official skepticism and foot-dragging. The aviators envisioned and promised great things, but their aircraft continually let them down when tested. Cunningham's troubles with his B–1 were all too typical of aircraft performance in this period. Even at their best, early naval airplanes, such as The Owl which the Marines took to Culebra in 1914, and which they did not use because the wings were deemed too weak to carry two men, had top speeds of no more than 50 miles per hour. Small fuel capacity and mechanical unreliability limited their range and endurance. The aviators' vision simply had outrun their technology, and until technology caught up, the opponents of aviation would hold strong ground.

In spite of administrative and technological frustrations, naval aviation achieved significant advances. The air station at Pensacola slowly acquired more men, aircraft, and equipment, and with these expanded its training and testing activities. Navy and Marine pilots practiced antisubmarine patrolling, bombing, and artillery spotting. Late in 1915, they began launching planes from an experimental catapult built on

A Curtiss C–3 being recovered by the USS North Carolina *during maneuvers off Pensacola, Fla., in July 1916. (National Archives Photo 80–G–426917).*

the Pensacola station ship, the cruiser USS *North Carolina.* In 1916, catapult experiments continued, and a few aircraft began operating on board warships of the fleet.

On 9 January 1915, the Marine contingent at Pensacola, now down to one flyer, McIlvain, was designated the "Marine Section, Navy Flying School." The section soon acquired two more pilots. Cunningham, who evidently had persuaded his wife to let him resume flying, reported to Pensacola in April for refresher training and flight duty. Early in the summer, the fourth Marine aviator, First Lieutenant Francis T. ("Cocky") Evans arrived and started flight training. The force of Marine enlisted mechanics also slowly increased, and at a still undetermined point in this period the Marines' first warrant officer aviator, Walter E. McCaughtry, learned to fly.*

In August 1915, as the result of an agreement

---

* McCaughtry was attached to naval aviation as early as June 1913 when a corporal. Apparently he learned to fly at some time during his tour of duty as an enlisted man. In June 1917 he was promoted to the temporary rank of captain and as an officer officially qualified as a naval aviator. He received permanent captain's rank in June 1920. (Pearson Comments)

between Secretary of the Navy Daniels and the Army Signal Corps, Navy and Marine pilots began training in land planes at the Signal Corps Aviation School in San Diego. Daniels had made this arrangement in the belief that defense of advance bases and, in the case of the Marines, possible joint operations with the Army, required an aviation force able to operate from either land or water. Lieutenant McIlvain was one of the first two naval aviators sent to the Army flight school. Cunningham followed him there in 1916. During this training he flew for the first time in a cockpit inside a fuselage instead of from a seat in the open in front of the wings of a primitive pusher. He wrote later that he would "never forget the feeling of security I felt to have a fuselage around me." [18] This training pattern persisted throughout the early period of Marine aviation. Marine pilots received basic flight instruction from the Navy and were designated naval aviators. Then they took land plane training at Army schools and advanced training with the Army and at their own airfields when they finally acquired them.

Besides learning to fly land planes, the Marine aviators participated in the aeronautic experiments at Pensacola, sometimes with near disastrous consequences for themselves. On 8 November 1916, for example, Cunningham attempted a takeoff from the catapult mounted on the *North Carolina*. His AB–2 seaplane overturned in the air, plunged into the water, and was wrecked but was towed to the ship and hoisted on board. Cunningham, although he seemed unhurt at the time, received a back injury which gave him months of pain. [19]

During 1916, with the European war continuing on its ever more destructive way, and the United States on the brink of war with Mexico and approaching a final confrontation with the Germans over U-boat depredations, the administration of President Woodrow Wilson began large-scale expansion of the Army and Navy. All branches of both services benefited, including Navy and Marine aviation. Urged on by the General Board, the Navy Department asked Congress for men, money, and aircraft. Plans took shape for a naval air arm of over 500 planes, and a series of interservice boards tried to define the respective roles and missions of Army and Navy aviation and began selecting sites for coastal airbases.

Congress, in the Naval Appropriations Act of 29 August 1916, provided $3,500,000 for aircraft and equipment. It also authorized the establishment of a permanent Naval Flying Corps of 150 officers and 350 enlisted men of the Navy and Marine Corps. Officers for this force could be appointed from warrant officers, enlisted men, or civilians, and were to be considered an addition to the legally authorized officer strength of the service. The act also authorized creation of a Naval Reserve, including a flying corps, recruited from former regular personnel or civilians, and it provided for a Marine Corps Reserve organized in the same branches as the Naval Reserve, thus by implication creating a Marine Corps Aviation Reserve. Opposition from the bureaus prevented creation of the Naval Flying Corps, but the reserve would grow rapidly after the American entry into World War I and would furnish most of the Navy and Marine pilots for the conflict.

While disagreements within the Navy Department blocked implementation of most of the personnel provisions of the act of August 1916, they did not prevent a rapid increase in and modernization of naval aviation's aircraft inventory. By the end of 1916, 60 new airplanes had been ordered, including 30 Curtiss N–9 seaplanes. In these machines, adapted from the JN, or "Jenny," trainer being built for the Army, naval aviators received their first "tractor" aircraft in which pilot and observer sat in cockpits in the fuselage with the engine and propeller in front of them. By the end of 1916, 25 of these airplanes, which were much safer* and more maneuverable than the old pushers, were in operation at Pensacola.

Early in 1917, while flying one of the new N–9s, "Cocky" Evans made a major contribution to American aviation safety. He did it largely by accident. He and other pilots at Pensacola had been arguing about whether one could loop a tractor-type seaplane. Evans and others insisted they could, while their opponents contended the heavy, fragile pontoons would make the maneuver impossible. On 13 February, at an altitude of about 3,500 feet above Pensacola on a routine flight, Evans decided to try to loop. His initial attempt failed, and his N–9 stalled and went into a spin. No American aviator up to this time had

---

* While the tractor-type biplanes were safer, their open cockpits, placed one behind the other, had their hazards for the man in the rear. This was particularly true if his pilot in the front seat enjoyed chewing tobacco, as many an early Marine aviator did. As an authority on Marine aviation puts it, a rear-cockpit gunner or observer with a tobacco-chewing pilot "had but a brief second to see the pilot's head start to swivel; then duck! The disposal of 'chaw' flew past at air speed in a fairly wet and scattered dispersion." (Gemeinhardt Comments)

worked out a method for recovering from a spin, and several had died in crashes as a result of this gap in their knowledge. Evans, apparently without realizing he was in a spin, instinctively pushed his control wheel forward to gain speed and controlled the turning motion with his rudder. Recovering from the spin, he kept trying to loop, stalling, spinning, and recovering until finally he managed to loop. To make sure he had witnesses for his feat, he flew over the seaplane hangars and repeated the whole performance. Not until then did he realize that besides proving a seaplane could loop he had solved a major safety problem. The aviators at Pensacola at once incorporated his spin-recovery technique in their training, and Evans was sent on a tour of military airfields to teach other pilots his method. Years later, on 10 June 1936, Evans received the Distinguished Flying Cross for this life-saving discovery.[20]

Marine Aviation Begins to Organize, 1917

At the end of 1916, out of a total of 59 commissioned officers and 431 enlisted men assigned to naval aviation, five of the officers and 18 of the enlisted men were Marines.[21] Marine aviation possessed no organization of its own beyond the amorphous "Marine Section" of the Naval Flying School, and it had no director or official spokesman at Headquarters. Its history up to this time had consisted largely of a series of individual exploits and disconnected episodes within the stream of naval aviation development.

Nevertheless, from the sending of Cunningham and Smith to Annapolis in 1912, the Marine Corps clearly had intended to build a distinct unit of its own attached to the Advance Base Force. By the end of 1916, the time for creation of such a unit seemed to be approaching. The Major General Commandant announced in his annual report for that year that a "Marine Corps Aviation Company" of 10 officers and 40 men would be organized "for duty with the Advance Base Force" at "as early a date as practicable." It would be equipped with both land- and seaplanes.[22]

The practicable date soon came. On 26 February 1917, Lieutenant Cunningham, soon to be promoted to captain, received orders to begin organizing at the Philadelphia Navy Yard an Aviation Company for the Advance Base Force.[23] Less than two months later, the United States declared war on the German Empire.

Lieutenant Alfred A. Cunningham standing in front of a Curtiss Pusher. (Photo from the Cunningham Papers).

CHAPTER II

MARINE AVIATION IN WORLD WAR I, 1917–1918

Marine Aviation Mobilizes

With the declaration of war against Germany, the Navy and Marine Corps entered a period of rapid expansion during which the air arms of both services grew in manpower and equipment and during which Marine aviation developed its own units and bases. After consultations with the Allies, the Navy Department adopted anti-submarine warfare as naval aviation's principal mission and began large-scale preparations for it. The office of Director of Naval Aviation quickly revived under the leadership of Captain Noble E. Irwin and veteran aviator Lieutenant Commander John Towers. With support from the Secretary of the Navy, Irwin and Towers effectively coordinated aviation activities in the Navy Department. Towers took charge of enrolling thousands of new officers and men in the Naval Aviation Reserve, and he set up training facilities for them at Army and Navy bases and universities to relieve swamped Pensacola. By the end of the war the manpower strength of naval aviation had reached over 6,700 officers and 30,000 enlisted men. In October 1917, the Navy Department adopted the "Seventeen Hundred Program" for building over 1,700 seaplanes of three different types, and to speed aircraft development and procurement it established the Naval Aircraft Factory at Philadelphia.

The Marine Corps, which entered the war with 511 officers and 13,214 enlisted men, began an expansion which would bring its strength to over 2,400 officers and 70,000 men on 11 November 1918. Under the energetic direction of Major General Commandant George Barnett, the Marines prepared to send a brigade to France to fight alongside the Army.

Marine aviation started an aggressive campaign to secure first its share of the manpower of the expanding Corps and then a chance to go to France and fight. In this campaign, Cunningham, commander-designate of the Aviation

Company, emerged as the principal leader and driving force. Although without a formally recognized office or title, he became *de facto* director of aviation for the Marine Corps. In 1917, he represented both Marine and naval aviation on the interservice board which selected sites for coastal air stations. He recruited men for Marine air units, sought missions for them to perform, and negotiated with the Navy, the Army, and eventually with the British for equipment and facilities. Looking back on this hectic time, Marine Major General Ford O. ("Tex") Rogers, whose own distinguished aviation career began in World War I, justifiably declared: "Cunningham was the father of [Marine] aviation, . . . absolutely, completely. Without him, there never would have been any aviation." [1] *

Marine aviation soon found itself split between two separate missions. The Aviation Company at Philadelphia, renamed the Marine Aeronautic Company and enlarged with men from the Aviation Section at Pensacola, from other Marine units, and from the recruit depots, was designated to fly seaplanes on anti-submarine patrols. During summer 1917, Major General Commandant Barnett secured Navy Department approval for the formation of a second Marine air unit of landplanes to provide reconnaissance and artillery spotting for the brigade being sent

---

* "Banney" Smith also distinguished himself in aviation at this time, but, following the pattern early established, he worked in naval aviation in general rather than Marine aviation. Ordered home from France in 1917, he directed much of the design and procurement of naval aircraft and then organized the aerial gunnery and bombing school at Miami. In 1918, he returned to Europe to organize the Intelligence and Planning Section for Naval Aviation at Navy Headquarters in Paris. After the war he had charge of assembling material and equipment for the famous transatlantic flight of the Navy NC-4s. Resigning his regular Marine commission in 1920, Smith entered the Marine Corps Reserve in 1937 and saw non-flying active service in World War II. He died in an automobile accident in 1946. (Biography Files, Reference Section, History and Museums Division).

to France. This unit, its organization patterned after that of Army aviation squadrons but with fewer men and machines, would consist of 11 officers and 178 men with six fighter planes, six reconnaissance aircraft, and four kite balloons for the artillery observers. Under an arrangement negotiated by Cunningham at Barnett's instructions, the Army Signal Corps would train pilots and crewmen for this unit and provide most of its aircraft and equipment.[2]

Marine aviation began a vigorous search for men for the projected units. As candidates for commissions flooded into the first wartime Marine officers' school at Quantico during the summer of 1917, Cunningham met them and preached the cause of aviation. Karl Day, a member of that first class who later rose to the rank of lieutenant general, recalled: "Major Rixey assembled the battalion and said Captain Cunningham at Headquarters had a message for us, and introduced Captain Cunningham. . . . He told us that we were going to have an aviation section, that we would go to France, and that he was down there to talk to anybody

who was interested in becoming a pilot." [3] Cunningham found plenty of volunteers. The officer candidates, many of them college athletes, responded to the challenge and glamor of aviation; as one of them put it, "It was a daring thing to do." Others, including Lawson H. M. Sanderson, who would become the Marine dive bombing pioneer, had other motives: "Well, hell, I thought I can ride better than I can walk. So I volunteered for aviation. . . . I'd only seen about two airplanes in my life, but I'd rather ride than walk." [4]

Out of dozens of volunteers from the first class at Quantico, Cunningham selected 18. Six of them eventually went into the Aeronautic Company for seaplane duty and the others joined the new landplane squadron. During the rest of the year, additional officers gradually expanded the ranks of Marine aviation. Few of these were regulars. Most were second lieutenants newly commissioned from civilian life and nominally members of the Marine Corps Reserve Flying Corps which had been authorized in the Naval Appropriation Act of 29 August 1916. The

The Kirkham tri-plane was typical of the many experimental models tried out by the Marines during aviation's early years. The Kirkham was made by Curtiss. (Marine Corps Photo 91590).

A JN-4B "Jenny." In 1917, Marines of the 1st Aviation Squadron trained in aircraft like this one (circa 1917). (Nat Archives RG 127-G Photo 517543).

Reserve, however, hardly had begun to organize when the war swamped it with new manpower, and in 1917, among the Marine aviators, "Nobody gave a damn and few, if any, knew who were regulars, temporaries, duration reserves, or what have you." [5]

By 14 October 1917, the Aeronautic Company had reached a strength of 34 officers and 330 enlisted men and had begun flight training, using two Curtiss R-6 seaplanes and a Farman landplane. On that date, the company was divided to form the two projected aviation units. The 1st Aeronautic Company (10 officers and 93 men) would prepare for seaplane missions while the 1st Aviation Squadron (24 officers and 237 men) would organize to support the Marine brigade being sent to France.

The 1st Aeronautic Company in the Azores

Of the two Marine air units, the 1st Aeronautic Company led the way into active service. In October, the company, commanded by "Cocky" Evans, now a captain, moved with its Curtiss R-6s to the Naval Air Station at Cape May, New Jersey, where it conducted seaplane training and coastal patrols. On 9 January 1918, enlarged to 12 officers and 133 enlisted men, the company embarked from Philadelphia for the Azores to begin anti-submarine operations.

For its anti-submarine mission, the company initially was equipped with 10 Curtiss R-6s and two N-9s. These were both single-engine, float-equipped, two-seater biplanes. The N-9 was the seaplane trainer with which the Marines had become familiar at Pensacola with a rear cockpit gun added and a more powerful Hispano-Suiza engine. The R-6, slightly larger than the N-9, had been purchased in great numbers by the Navy under the "Seventeen Hundred" Program. The company later received six Curtiss HS-2L flying boats. Each of these patrol planes carried a crew of two and with its single 330-horsepower Liberty engine could achieve a top speed of about 90 miles per hour. With a cruising range of about 400 miles, the HS-2L represented a marked improvement over the R-6 and the N-9, and it greatly enhanced the company's ability to carry out its mission. [6]

Throughout 1918, the Aeronautic Company maintained its vigil from its base at Punta Delgada on the island of San Miguel. The assignment had its pleasant aspects. "There was wonderful flying weather, people were very friendly . . . They took us right into their homes and did everything they could for us, so it was good to be there." Operating within a 70-mile radius of the islands, the company, the first American aviation unit of the war to deploy with a specific mission, flew regular patrols to deny enemy submarines ready access to the convoy routes and deprive them of a safe haven in the Azores. An enlisted gunner and ground crewman of the force summed up the results: "We saw a few out there; in fact we dropped a few bombs, but as far as we know we didn't damage anything . . . But we kept them submerged, I think." [7]

An HS–2L flying boat of the type Marines used in the Azores during 1918. (Nat Archives RC 127–G Photo 517536).

HS–2L in flight in 1918. These aircraft greatly increased the anti-submarine effectiveness of the 1st Aeronautic Company. (Nat Archives RG 127–G Photo 517529).

Beginnings of the 1st Marine Aviation Force

The Marine landplane unit, the 1st Aviation Squadron under Captain "Mac" McIlvain, followed a more tortuous road to war. Under an agreement reached on 10 October between Captain Cunningham and Colonel Henry H. Arnold of the Army Signal Corps, the squadron was to receive basic flight training at the Army Aviation School at Hazlehurst (later Roosevelt) Field at Mineola, Long Island. Then it would move to the Army school of advanced flying at Houston, Texas. After this training was completed, "The Squadron will . . . be ready for service in France and the Army will completely equip it with the same technical equipment furnished their squadrons." [8]

Initially, the squadron's development went according to plan. On 17 October, the unit moved from Philadelphia to Mineola to begin training, and early in November the six officers of its balloon contingent were ordered to Fort Omaha, Nebraska, for instruction as aerial observers. The main body of the squadron at Mineola lived in tents near the runways and flew JN–4B "Jenny" trainers with civilian instructors, some of whom proved less than satisfactory. One of them, a Marine remembered, "was scared to death. He wouldn't let anybody touch the controls. I had four or five rides with him, and he never once let me touch the throttle, the wheel, or touch the rudder. So we raised hell about him, and he got fired." [9]

While McIlvain's squadron trained at Mineola, a third Marine aviation unit began forming at Philadelphia. This was the Aeronautic Detachment, organized on 15 December under Captain Roy S. Geiger, Marine Aviator Number 5, who had qualified early in 1917, with four officers and 36 enlisted men, most of them drawn from the 1st Aviation Squadron. This unit's mission remained uncertain at the time of its formation, but it apparently was intended for work with the Advanced Base Force. [10]

While the Marine land-based aviation force began organizing and training, Captain Cunningham sought a firmly defined mission for it. During November and December of 1917, he toured the Allied aviation facilities in France, visiting French and British air bases and flight schools and flying with the Allies on missions over the German lines. [11]

During his trip, Cunningham strove to persuade the Army to attach the Marine landplane squadron to the Marine brigade in France as originally had been intended. However, in Cunningham's words, the Army aviation authorities "stated candidly that if the [Marine] squadron ever got to France it would be used to furnish personnel to run one of their training fields, but that this was as near the front as it would ever get." [12]

With the intended mission of his force thus wiped out, Cunningham turned to enlarging the Navy's air role in France. Navy planners initially had envisioned conducting only anti-submarine patrols with seaplanes. Cunningham, however, in conferences with U.S. Navy officers at Dunkirk and with officers of the British destroyer patrol, discovered a need for bombers to attack the German submarines which operated from bases at Zeebrugge, Bruges, and Ostend on the Belgian coast. Such attacks could help to stem the submarine onslaught that early in 1918 still threatened to cut Britain's oceanic supply lines. Cunningham prepared a plan for a land-based force of Marine and Navy planes to take over this mission, which the British lacked the aircraft and pilots to perform. On 5 February 1918, with strong support from U.S. Navy officers in Europe and from the Allied authorities, he presented his plan to the General Board of the Navy. The board approved the plan and ordered the formation of a Northern Bombing Group to carry it out. On 11 March, Cunningham received orders to organize and take command of a 1st Marine Aviation Force which would be formed by combining Geiger's and McIlvain's detachments at Miami. This force would constitute the Marine element of the new bombardment group. [13]

After the initial decision to form it, the Northern Bombing Group went through several changes of mission and equipment. These changes resulted from debates between the Navy Department, Rear Admiral William S. Sims in London, the War Department, and the Allies. From bombing U-boats in the shallow coastal waters, the group's mission changed to bombing the German submarine pens in the Belgian ports. From flying fighters to escort the bombers, the Marine squadrons' role changed to conducting daylight bombing raids, using the British-designed DH–4. The Navy wing of the force, flying large Italian-built Caproni bombers, would carry out night raids. [14]

While Cunningham was seeking a mission for his Marine aviation force, the units from which it was to be created had been undergoing adventures of their own. McIlvain's 1st Aviation Squadron remained at Mineola until 1 January

Two Curtiss R–6s (foreground) and an N–9 of the 1st Aeronautic Company at Punta Delgada, Azores, 1918. (Nat Archives RG 127–G Photo 529925).

1918, by which date temperatures had reached 16 degrees below zero, rendering flight training almost impossible, and creating a threat to the health of the Marines, who still were living in tents. When the squadron medical officer declared a change of camp necessary for reasons of health, Captain McIlvain loaded his men, equipment, and aircraft on a train and headed southward.

The aviators left Mineola with little advance planning and, according to some accounts, without orders specifying their new station.* The squadron soon received instructions to report to

the Army's Gerstner Field at Lake Charles, Louisiana, but when the Marines arrived there the base commander refused to take them in because he had no authorization to do so from the War Department. For a day or so, the Marine aviators lived on board their train and ate in a borrowed Army mess hall. Then the necessary orders arrived, and the base commander allotted the Marines quarters and incorporated them into the landplane training program. Even then, the Marines had to uncrate, assemble, and test fly their own trainers before they could take to the air. Gradually the confusion sorted itself out and the Marines settled down to work, flying stick-controlled JN–4D trainers instead of the wheel-controlled JN–4Bs they had used at Mineola. They also practiced with the rotary-engine S4–C Thomas-Morse Scout.**

---

* An order exists, dated 31 December 1917, directing the squadron to move to its eventual new station, Lake Charles, Louisiana; but participants recall leaving Mineola without orders and stopping in Washington, D.C., on the way south while officers went into town to HQMC to ask what they should do next and received only suggestions they keep heading southward. At some time during the trip the squadron did receive orders to Lake Charles, but possibly the surviving copies were back-dated.[15]

** In a rotary engine, the engine rotated around the crankshaft, rather than the engine standing still and the crankshaft rotating as in the more common radial engine.

Captain Geiger's Aeronautical Detachment, the second component of the planned Marine bombing force, also moved early in 1918. On 4 February, Geiger received orders to take his detachment, which now consisted of 11 officers and 41 men, to the Naval Air Station at Miami, Florida. The unit left Philadelphia three days later.[16] Soon after arriving in Miami, Geiger, now seeking a base for the entire 1st Aviation Force, moved his command from the main Navy field at Coconut Grove near Miami to a small, sandy airstrip on the edge of the Everglades, which was owned at the time by the Curtiss Flying School.

To secure Marine training facilities independent of the Army, Geiger absorbed the entire Curtiss Flying School into the Marine Corps, arranging to commission the instructors in the Reserves and to requisition the school's Jennies. Cunningham cleared the way for this unorthodox action and also obtained for Geiger 20 Jenny land planes for use as trainers. On 1 April, McIlvain's squadron arrived at the Miami field from Lake Charles, at last consolidating at one location the nucleus of the 1st Aviation Force.

Cunningham, still serving as *de facto* Director of Marine Aviation as well as commanding the 1st Aviation Force, launched a campaign of improvisation to bring his squadrons to full strength in men and machines. He again visited the officers' school at Quantico and obtained six more volunteers whom he sent to Miami. He collected other volunteers elsewhere, men whom one Marine pilot referred to as "strays that Cunningham . . . picked up. I don't know where he got them." [17] Geiger recruited some of the civilian students at the Curtiss Flying School, promising them commissions if they satisfactorily completed pilot training. During March and April, 18 newly commissioned Marine lieutenants arrived in camp. Enlisted ground crewmen also appeared, some of them skilled mechanics, electricians, carpenters, and blacksmiths, others "just good Marines who had little more than basic military training." [18]

Even with this re-enforcement, Geiger's and McIlvain's detachments combined could not furnish enough pilots for the planned four squadrons. Realizing this, Cunningham toured the Navy air installations at Pensacola and Key

While the Standard E–1 was basically an Army aircraft, several were flown by Marines while training at Army airfields in 1917–1918. At the controls is "Curley" Newman while "Red" Weiler looks on. (Photo from the Goodyear Kirkham collection).

DH–4B of the 1st Aviation Force with the force's insignia of a Marine Corps Globe and Anchor superimposed on the Allied rondel. All DH–4Bs of the force had identification numbers with a "D" prefix. (Photo from Alfred A. Cunningham Papers.)

Around 1918 tents were used as hangars for Marine aircraft based at the Marine Flying Field near Miami, Fla. The planes are JN–4s. (Photo from the Goodyear Kirkham collection).

West and recruited naval aviators, most of them young reservists who wanted to go to France. These officers, already qualified Navy seaplane pilots, disenrolled from the Navy, enrolled in the Marine Corps, and reported to the Marine field at Miami for landplane training. Of 135 aviators who eventually flew in France with the Marine bombing force, 78 were transferred Navy officers.[19]

The Marines at Miami adopted an intensive training schedule, sandwiching into a few weeks basic flight instruction in seaplanes (necessary to qualify everyone for Navy wings), elementary landplane training, formation flying, aerobatics, and the rudiments of aerial tactics, gunnery, bombing, and reconnaissance. Some of the enlisted men were detailed and trained as air gunners and observers. Others took instruction, usually on the job, as mechanics, armorers, and ground crewmen. Officers and men worked from daylight until dark under less than ideal conditions. Drifting sand and dust filtered into engines, increasing maintenance difficulties, and the swamps of the Everglades which bordered the field made every forced landing into a major rescue and recovery problem. Haste and overwork took their inevitable toll. The force lost four officers and three enlisted men killed in accidents.

On 16 June 1918, Cunningham organized a headquarters detachment and four squadrons, designated A, B, C, and D. Geiger and McIlvain commanded A and B respectively; Captain Douglas B. Roben commanded C, while First Lieutenant Russell A. Presley commanded Squadron D. The four squadron commanders at once left for France, where they selected two airfield sites and established liaison with the Northern Bombing Group. On 10 July, the 1st Aviation Force received orders to embark for the front. At this time, a British aviator sent to appraise the squadrons' state of readiness pronounced them fit for combat, but a Marine aviator who was there had a different view:

> We had flown nothing but Jennies. We got one DH–4 [the bombing plane they were to fly in France], and all of us in Miami got one flight in the first DH–4. . . . We had one flight. Our gunnery training had consisted of getting into the rear seat and using a Lewis gun, shooting the targets on the ground. None of us had ever fired a fixed gun in our lives. None of us had ever dropped a bomb in our lives. . . .[20]

Whether ready or not, most of the personnel of the 1st Aviation Force headed for France in the expectation that their aircraft and equipment would be waiting there for them when they arrived. On 13 July, the force, less Squadron D which remained behind temporarily, boarded trains at Miami. On the way to New York, their port of embarkation, they stopped over at Philadelphia to receive an official band-accompanied farewell from the city which had strong claims to the title of birthplace of Marine aviation. On 18 July, the 107 officers and 654 enlisted men of the three squadrons sailed from New York for France on board the transport USS *De Kalb*.

Marine Aviation Expands

The 1st Aviation Force left behind it a Marine air arm that was emerging rapidly from confused improvisation into steady, businesslike expansion. At Miami, the Marine Flying Field, as the Curtiss Field had been renamed, had grown from a sleepy airstrip bordered by a couple of wooden-framed hangars into a bustling military complex of hangars, storehouses, machine

DH–4s on a flight line in France in 1918. The planes are taking off for a bombing raid on German lines. (Photo from the Goodyear Kirkham collection).

shops, tent camps, and gunnery and bombing ranges. The field continued in operation after the 1st Aviation Force left, first to complete the formation of Squadron D and then to train additional aviation personnel and to serve as the base for Marine air patrols of the Florida coast. In Washington, Captain Harvey B. Mims, who kept in close touch with Cunningham in France, acted as director of Marine aviation.[21]

During 1918, the authorized strength of Marine aviation was set at 1,500 officers and 6,000 men. To reach this number of personnel, Marine aviation, besides recruiting more officer pilots, began training enlisted aviators. The first class of 25 candidates entered this program on 10 July 1918.[22] These Marines, who had to meet special educational and physical qualifications,* received the temporary rank of gunnery sergeant. They first attended a 10-week academic course at the Massachusetts Institute of Technology. From there they went to Miami for flight training. Candidates who successfully completed flying school received commissions as second lieutenants in the Marine Reserve Flying Corps. At Great Lakes Training Station in Illinois, a Marine section of the Navy Mechanics' School prepared enlisted Marines for ground duty in aviation, as did a similar school in San Diego.[23]

The 1st Marine Aviation Force in France

While Marine aviation expanded in the United States, its vanguard in the war zone, the 1st Aviation Force, disembarked at Brest on 30 July. Administrative and supply problems dogged Cunningham and his Marine aviators from the day of their landing. These problems were compounded by a lack of co-ordination and firm understanding between Cunningham's headquarters and that of Captain David Hanrahan, USN, commanding officer of the Northern Bombing Group.[24]

Upon landing at Brest, Cunningham found that no arrangements had been made to move his squadrons the 400 miles to their selected base locations near Calais. Cunningham solved this problem by requisitioning a French train for the two-day trip. When he sent a working party to the Navy supply base at Pauillac, 30 miles

from Bordeaux, to collect the force's motor vehicles, the officer in charge discovered that "All our . . . trucks . . . had gotten mixed up and gotten into the Army pool, and I had to go down there, drag em out of that pool, and find drivers, and send those things North."[25]

After they reached Calais, the Marines, billetted temporarily in a British rest camp, began work at the landing field sites selected by their advance party. Squadrons A and B were located at Oye, a town between Calais and Dunkirk. Squadron C occupied a site at LaFresne, 12 kilometers southwest of Calais, while Cunningham established his headquarters at the town of Bois en Ardres.

Cunningham then discovered that he was not going to have any aircraft for a while. Before leaving for France, he had secured from the Army 72 DH–4 bombers. These British-designed machines, part of a large number being manufactured in the United States, would be shipped to France and assembled there for issue to the Marines. The planes duly arrived at Pauiliac at about the time the 1st Aviation Force disembarked at Brest, but due to delays in assembly, the first one did not reach the squadrons until 7 September, and Cunningham learned that most of them by some administrative oversight had been shipped to England. Cunningham "in desparation" struck a bargain with the British. They had a surplus of air frames for the DH–9A, a modification of the DH–4, but no engines; the Americans had in Europe a surplus of the Liberty engines for which the DH–9A was designed but few airplanes in which to put them. Cunningham, with the approval of U.S. Navy authorities, agreed with the Royal Air Force that for every three Liberty engines delivered to the British, they would return one to the Marines mounted in a completely equipped DH–9A. By this means and as the result of the eventual delivery of some of his Army machines, Cunningham by the time of the Armistice had secured 36 airplanes, about half of his force's planned strength. Of these, 20 were DH–9As and 16 were DH–4s.[26]

Unable to get his fliers into the air at once in American machines, Cunningham again turned to the British.** He knew that the RAF had an abundance of aircraft, but a shortage of pilots as a result of years of combat losses. Within nine

---

* Candidates had to be enlisted Marines, have a "superior" physique, and weigh between 135 and 165 pounds. Age limits were 19 to 39 years. Candidates had to have completed at least two years of college.

** The Marines were stationed behind the sector of the front held by the British armies rather than in the American sector which was to the south and east of them.

The first aerial resupply mission in Marine aviation history occurred 2–3 October 1918 when Marine Captain Robert S. Lytle (right) and Gunnery Sergeant Amil Wiman helped airdrop 2,600 pounds of food and stores to a French infantry regiment. Also taking part in the action were Marine Captain Francis P. Mulcahy and Gunnery Sergeant Thomas L. McCullough. (Photo from the Goodyear Kirkham collection).

days of the 1st Aviation Force's arrival in France, he had arranged for Marine pilots to fly bombing missions with RAF Squadrons 217 and 218, which operated DH–4s and DH–9s, the same types of aircraft the Marines were to receive. Soon Marine pilots, many of whom added British overseas caps and canes to their uniforms, were flying combat missions over the German lines. They served with the British in rotation, so that every Marine aviator would go on at least three raids.

The Marines now completed their training under fire and under the tutelage of veterans whom they came to admire and respect. Karl Day, who flew with Squadron 218, a mixed lot of men from all the British dominions, said of the outfit's commander, Major Bert Wemp, a Canadian: "He taught me what it means to be an officer and a gentleman. He was a remarkable commanding officer." The British on their side welcomed the American reinforcements but treated them on missions with grim realism.

"They put us—always the newcomers were the last on the right, in the 'V', because if you got shot you hadn't lost anything." [27]

These joint operations with the British produced some notable Marine achievements. On 28 September, while flying with Squadron 218, First Lieutenant Everett S. Brewer and Gunnery Sergeant Harry B. Wersheiner shot down the Marines' first enemy aircraft in a dogfight over Courtemarke, Belgium. Both Marines were severely wounded in the engagement. On 2 and 3 October, Marine airmen, also operating with Squadron 218, flew the first aerial resupply mission in the history of Marine aviation. On those days, Captain Francis P. Mulcahy and his observer, Gunnery Sergeant Thomas L. McCullough, and Captain Robert S. Lytle and his observer, Gunnery Sergeant Amil Wiman, flew through heavy German fire to drop over 2,600 pounds of food and stores to a French infantry regiment isolated by mud and surrounded by enemy near Stadenburg.

While the aircrews flew with the British, the rest of the Aviation Force worked on their flying fields. Without engineering equipment, each squadron had to build its own runways, hangars, living quarters, and other base facilities. The Marines dug sugar beet plants out of the flat fields with shovels and then levelled their airstrips with a borrowed Navy steamroller. They secured from the British large amounts of construction material, including portable canvas and wood hangars. Housing for pilots and crews went through three stages: tents with dirt floors, tents with wooden frames and floors, and finally portable wooden buildings boasting stoves, cots, and furniture made from shipping crates. In spite of his men's efforts, however, Cunningham in November did not consider them well enough housed to stay in their camps through the winter.[28]

On 5 October, Squadron D arrived at La-Fresne with 42 officers and 183 enlisted men, raising the strength of the entire Force to 149 officers and 842 rank and file. The Marines now redesignated their squadrons as Numbers 7, 8, 9, and 10 to conform to the Northern Bombing Group's identification system.

The mission of the force also changed at this time. Under pressure from the Allied offensives then in progress, the Germans evacuated their submarine bases on the Channel coast, eliminating the Marines' planned mission. Attached to the RAF, the Marine squadrons now shifted to general support of the British and Belgian armies, which were pressing their final assault against the crumbling German lines.

By 12 October, the Marine squadrons had received enough of their own DH–4s and DH–9As to begin flying missions independently of the British.[29] Their DH–4s, similar to those they had been flying in combat with the RAF, were versions of one of the more successful World War 1 aircraft. First flown in 1916, this British-designed two-seater biplane had a wingspread of 42½ feet and a length of 30½ feet. It was built of wood and fabric like other aircraft of the time, but had the front part of its fuselage covered with plywood. The model used by the Marines, which was fitted with a 400-horsepower American Liberty engine, could achieve a top speed of about 125 miles per hour and had a ceiling of 17,500 feet. It had a range of 270 miles and could climb to 10,000 feet in 14 minutes. Armed with four .30-caliber machine guns (two fired by the pilot through the propeller and two on a ring mount fired by the observer) and able to carry

460 pounds of bombs in wing racks, the DH–4 had enough speed, maneuverability, and fire power to hold its own against contemporary fighters. With its easily ruptured, pressurized fuel tank located between the pilot and the observer, the DH–4 received the ominous nickname of "The Flaming Coffin," less from a tendency to burn when hit by enemy fire than from the likelihood of its catching fire during otherwise minor noncombat mishaps.

With the British, the Marines had flown the DH–9, and as a result of Cunningham's negotiations their own squadrons received DH–9As as well as DH–4s. The DH–9, developed in 1917 by the makers of the DH–4, had been intended as an improvement upon the earlier machine. It was identical in construction and dimensions to the DH–4, but had its fuel tank located in the fuselage ahead of the pilot's cockpit, which was placed further toward the tail and closer to the observer's position. This allowed the crew to communicate more easily with each other in flight. However, the early DH–9s, underpowered and plagued by engine trouble, proved much inferior in performance and reliability to the DH–4. Accordingly, when the American Liberty engine became available late in 1917, the British modified the DH–9 for it, producing the DH–9A. The new version had a wider wingspread than

Second Lieutenant Ralph Talbot and his observer, Corporal Robert G. Robinson, earned the Navy Medal of Honor for their actions in fighting off 12 German aircraft on 14 October 1918.

The colors of the 1st Marine Aviation Force being presented by Mme. Trouille, wife of the Mayor of Ardres, to 2d Lieutenant William E. Russell, USCMR, Force Intelligence Officer, 27 November 1918. (Photo from the Alfred A. Cunningham Papers).

either the DH–4 or the DH–9, an enlarged radiator for the 400-horsepower Liberty, and a more strongly framed and braced fuselage. This was the aircraft Cunningham obtained. For all practical purposes, it was identical in performance, range, bomb load, and armament to the DH–4, and like the DH–4 it was manufactured under license by United States aircraft firms.[30]

On 14 October 1918, Captain Robert S. Lytle of Squadron 9 led the Marines' first mission in their own aircraft. With a flight of five DH–4s and three DH–9As, he struck the German-held railroad yards at Thielt, Belgium, dropping 2,218 pounds of bombs.

The bombings met no serious resistance, but on the way back to base, 12 German fighters (eight Fokker D–VIIs and four Pfalz D–IIIs) intercepted the Marine formation. In the ensuing melee, the Germans, following their usual tactics when fighting bombers, concentrated their attack on one machine, the DH–4 piloted by Second Lieutenant Ralph Talbot, one of the Naval Reserve officers who had transferred to Marine aviation. Talbot's observer, Corporal Robert G. Robinson, an expert gunner, quickly shot down one attacker, but two others closed in from below, spraying the DH–4 with bullets, one of which shattered Robinson's elbow. After clearing a jammed machine gun, Robinson continued to fire until hit twice more, while Talbot frantically maneuvered for advantage. With Robinson unconscious in the rear seat, Talbot brought down a second German with his front guns, then put the plane into a long dive to escape the rest of the enemy. Crossing the German lines at an altitude of about 50 feet, he landed safely at a Belgian airfield, from which Robinson was taken to a field hospital. He recovered from his wounds. For this exploit, Talbot and Robinson both received the Medal of Honor.

Captain Lytle also had a narrow escape. While he was trying to maneuver to aid Talbot, his engine failed; he glided back to the lines and brought his plane down immediately in front of the Belgian forward positions. Lytle and his observer scrambled out of the plane into the relative security of the trenches, and later that night Marine ground crews dismantled the aircraft and brought it back to base. The rest of the Marine formation returned safely to their own fields.[31]

Between 14 October and 11 November, the Marines carried out a total of 14 raids. They bombed railway yards, canals, supply dumps, and airfields. Always flying without fighter escort, they frequently braved German antiaircraft

On 22 October 1918, the first Marine aircraft was destroyed as a result of enemy action when seven German aircraft attacked and shot down a plane flown by 2d Lieutenant Harvey G. Norman. Norman and his observer, 2d Lieutenant Caleb C. Taylor, were killed in the crash.

fire and had several more clashes with German planes. In one of these, on 22 October, seven enemy fighters separated the craft piloted by Second Lieutenant Harvey G. Norman from the formation and shot it down, killing both Norman and his observer, Second Lieutenant Caleb W. Taylor. This was the first Marine aircraft lost to enemy action.

During their tour of duty in France between 9 August and 11 November, the Marines of the 1st Aviation Force took part in 43 missions with the British, besides launching their own 14 raids. According to later estimates, they dropped 15,140 pounds of bombs while flying with the British and 18,792 pounds of bombs on their own missions. At a cost of four pilots killed and one pilot and two gunners wounded, they scored four confirmed kills of German fighters and claimed eight more. In addition to combat casualties, the Marines lost Lieutenant Talbot killed on a test flight and four other officers and 21 enlisted men dead in an influenza epidemic which for a while in October paralyzed operations. During their brief period in combat, the Marines of the 1st Aviation Force won a total of 30 awards and decorations, including Talbot's and Robinson's Medals of Honor and four Distinguished Service Medals.[32]

Even before the signing of the Armistice,

Cunningham requested the early return of the 1st Aviation Force to the United States. He did this partly to prevent his ill-housed men from having to spend the winter in Belgium and Northern France and partly because, in his words, "I think we could accomplish much more at home, getting our Aviation service established under the new conditions of peace." [33] To the disappointment of some of his men, who had hoped to see Paris and Germany before their adventure ended, Cunningham obtained what he wanted. The 1st Aviation Force sailed for home on the USS *Mercury* in December 1918.

Marine aviation during the war had grown from a few men and machines into an organized branch of the Corps, with its own bases and training programs. It had proven itself in combat. Now it would have to prove itself in peacetime to Congress, the American public, and the rest of the Marine Corps.

CHAPTER III

ADVANCE TOWARD MATURITY, 1919–1929

Battle for Survival, 1919–1920

Major Cunningham (he had received a wartime promotion to temporary rank on 28 August 1918) returned home with the 1st Aviation Force and resumed his position as officer in charge of aviation. At this time, the Marine air arm which he had done so much to create contained 282 officers and 2,180 enlisted men, counting the units in the Azores and Miami as well as the Marine squadrons in France.[1]

Postwar demobilization began immediately, for the Marine air arm as for the other services. The 1st Aviation Force disbanded at Miami in February 1919, and most of the remaining Marine air personnel and equipment were dispersed to Parris Island and Quantico. From the remnants of his establishment, Cunningham formed a new Squadron D, which went to the Dominican Republic to support the 2d Provisional Brigade stationed there, and Squadron E, which deployed to Haiti to support the 1st Provisional Brigade. In September 1919, the Marine Flying Field at Miami was closed down.[2]

The Marine Corps, along with the other services, began a desperate struggle to persuade Congress to maintain at least their prewar personnel strength with the required bases, facilities, and equipment. Within this larger battle over appropriations and legislation, Major Cunningham fought for permanent status for Marine aviation. He labored under the disadvantage that the Marine air arm had no permanent bases or facilities and no precedent for peacetime strength or organization. Cunningham also realized that he would face opposition from within the Marine Corps. He summed up the nature and reasons for this opposition: "One of the greatest handicaps which Marine Corps aviation must now overcome is a combination of doubt as to usefulness, lack of sympathy, and a feeling on the part of some line officers that aviators and aviation enlisted men are not real Marines." Much of this attitude within the Marine Corps, as Cunningham pointed out, resulted from the fact that the Marine air squadrons in France, through no fault of their own, had not been allowed to support the Marine brigade.[3]

In an appearance before the General Board on 7 April 1919 and through an article published in the *Marine Corps Gazette* in September 1920, Cunningham sought to persuade skeptics within and outside the Marine Corps of the value of aviation. In these presentations, he defined what is still the primary mission of Marine aviation and anticipated the development of the modern air-ground team. He started his argument with the assumption, as he put it in his *Gazette* article, "that the only excuse for aviation in any service is its usefulness in assisting the troops on the ground to successfully carry out their operations." For Marine aviation, this would mean primarily support of the advance base and expeditionary forces in seizing and holding assigned objectives.

In his *Gazette* article, the principal published expression of his concepts of Marine aviation, Cunningham outlined in detail the possible roles of Marine air squadrons in supporting an opposed beach landing. Anticipating features of World War II operations, he discussed aerial reconnaissance and suggested that bombers could isolate the landing area by attacking railroads, roads, and enemy re-inforcements. During the actual landing, he wrote, "the planes could with machine gun fire and small fragmentation bombs so demoralize resistance as to make the task of landing much easier and safer." He emphasized the need for close and constant communication between air and ground units and pointed to the usefulness of radio in maintaining such communications. He also expanded on the ways in which aviation could aid the Marines in operations against Central American guerrillas. Marine fliers during the 1920s would illustrate these passages of his article by their actions.[4]

A cross-country "Jenny," based at the Marine Flying Field, Miami, Fla., about 1922. (Photo from Goodyear Kirkham collection).

A Thomas-Morse (MB–3) postwar fighter. In 1922, 11 MB–3s were transferred from the Navy to the Marines. (Marine Corps Photo 717548).

The DH–4B appeared in many guises, including this O2B–1 of VO–3M (Observation Squadron 3) with the 1st Aviation Group in 1921, which served the Marines as an observation and attack plane. (Nat Archives RG 127–G Photo A402979).

Arrow indicates a Fokker D–7, one of two D–7s to fly with the Marine Corps. The D–7s were captured during World War I. This inspection lineup including VE–7s and MB–3s is at Quantico in 1922. (Photo courtesy Goodyear Kirkham collection).

As a result of the efforts of Cunningham and others, Marine aviation won its battle for survival. After an 18-month legislative struggle, Congress established the Marine Corps at approximately one-fifth the manpower strength of the Navy, or 26,380 men. It then authorized an additional 1,020 Marines for aviation, bringing the total Marine force to 27,400. Along with its manpower, Marine aviation soon received permanent bases. By the end of 1920, air stations were under construction at Quantico and Parris Island, and the establishment of a field at San Diego had been approved. The Quantico and Parris Island installations would serve as bases for the air elements of the Atlantic coast expeditionary force, as well as centers for advanced flight and tactical training. The planned San Diego field would perform the same functions for the air arm of the Pacific coast expeditionary force.[5]

Organization and Mission, 1920–1929

With its manpower and bases assured, Marine aviation could establish a stable organization.[6] At Headquarters Marine Corps, Cunningham's position received formal recognition in January 1919, when he was assigned to duty as head of the Marine section of naval aviation. This section had charge of the recruitment and training of Marine air personnel and acted as the channel for aviation supply requisitions to both the Navy and the Marine Corps. Because Marine aviation continued to draw heavily upon the Navy for aircraft, supplies, and training facilities and because Marine air squadrons would operate in close co-operation with naval units, Cunningham and his successors were placed under the authority of both the Major General Commandant and the Director of Naval Aviation. The latter official in turn was part of the office of the Chief of Naval Operations. As Cunningham described it in March 1920:

> This office is so closely associated with the office of the Director of Naval Aviation as to be for all practical purposes a part of that office. In other words, it is a closely connecting link between the Major General Commandant's office and the Director of Naval Aviation regarding purely aviation matters and handles all Marine Corps matters which refer to aviation. This arrangement is working very satisfactorily and it is recommended that it remain in force. . . .[7]

Within Headquarters Marine Corps, the Aviation Section on 1 December 1920 was put under the control of the newly formed Division of Operations and Training. This office would oversee the materiel and personnel of the air arm and would direct the joint training of aviation and ground forces. This reassignment of the Aviation Section thus promoted closer ties between Marine aviation and the rest of the Corps, but the effectiveness of the arrangement depended heavily on the interest in aviation of the Directors of Operations and Training.

Major Cunningham remained at the head of the Aviation Section until December 1920, when he left the office as the result of an unusual set of circumstances. Cunningham in 1920 was not the senior Marine aviator in rank. That distinction belonged to Lieutenant Colonel Thomas C. Turner, who had entered the Marine Corps as a second lieutenant in 1902 and reached the permanent rank of major by 1917.[*] An aviation enthusiast like Cunningham, Turner learned to fly in his spare time while commanding the San Diego Marine barracks. With Marine Corps permission, he took his flight training at the Army Signal Corps Aviation School at San Diego, and when the United States entered World War I, he requested and received an assignment to aviation duty with the Army. He served with the Army at Ellington Field, Texas, throughout the war and also on 14 March 1918 received his wings as a naval aviator. Cunningham, who regarded Turner as a rival, kept him out of Marine aviation. Returning to the Marine Corps in 1919, Turner served with ground troops in Haiti, where he was cited for valor during a bandit attack on Port au Prince. In 1920, he requested aviation duty with the Marines, and Major General Commandant John A. Lejeune, a friend of Turner's, recognized his seniority and made him head of aviation. Cunningham received command of the squadron in the Dominican Republic.[8]

Cunningham commanded the squadron in Santo Domingo for a year and a half. Then, under a policy which he himself strongly advocated of returning Marine fliers to ground duty after five-year tours in aviation, he was transferred to a series of non-flying staff billets. In 1928, he requested a return to aviation duty but was turned down on grounds that no administrative posts were open and that he was too old to

---

[*] Cunningham did not receive his permanent promotion to major until 4 March 1921.

resume flying. He retired in 1935 and died four years later. His contribution to Marine aviation is best summed up in his own words, written in 1928 to Major General Commandant Lejeune:

> I was the first Marine officer to fly and spent the best years of my career working with enthusiasm to advance Marine Corps aviation. I did the unappreciated pioneering work and stuck by it during the time when no one considered it important enough to be desirable duty, paying the usual toll which pioneering demands. I was in charge of Marine Corps aviation during its first ten years and built it, mostly under the handicap of lack of interest in all aviation, from one officer to 300 trained pilots and about 3,000 mechanics, personally arranging for the details of personnel, material, training and organization.[9]

Whatever personal differences he may have had with Cunningham, Turner shared his predecessor's devotion to the interests of Marine aviation. As head of the Aviation Section, he aggressively continued Cunningham's work of building up the air service, personally participating in long-distance flights to demonstrate aviation's capabilities. On 2 March 1925, he turned the Aviation Section over to Major Edwin H. ("Chief") Brainard, still another shrewd and determined advocate of Marine aviation. Brainard during his tenure emphasized the recruitment of more officer pilots and directed the revival and expansion of the Marine Aviation Reserve.[10] In 1929, when Brainard left the Marine Corps for private industry, Turner, who had finished a tour in command of the squadrons then serving in China, resumed his post as head of the Aviation Section.

Throughout the decade, the office which directed the efforts of Marine aviation remained modest in size. Captain Louis E. Woods, who reported there for duty in 1926, found that the section then consisted of Major Brainard, Major Ford O. ("Tex") Rogers, and himself plus three civilian clerks. Under Brainard's overall direction, Wood "had more to do with personnel and training and so forth . . . and Tex had more of the materiel end." [11]

While Cunningham, Turner, and Brainard represented Marine aviation in Washington, more or less permanent operating organizations took shape in the field. On 30 October 1920, Major General Commandant Lejeune approved an air table of organization which provided for aircraft "wings," each of which was to be composed of two to four squadrons which in turn were divided into "flights." The existing aircraft and personnel were formed into four squadrons, each of two flights. The 1st Squadron (Flights A and B) consisted of the planes and crews in the

Lieutenant Colonel Thomas C. Turner, chief of Marine Aviation from 1920–1925 and 1929–1931. (Photo from the Goodyear Kirkham collection).

Dominican Republic. The 2d and 3d Squadrons (Flights C, D, E, and F) were stationed at Quantico, which throughout the decade contained the largest concentration of Marine aircraft and aviation personnel. The 4th Squadron (Flights G and H) was based at Port au Prince, Haiti, to support the 1st Provisional Brigade. The detachment at Parris Island, designated Flight L, was under orders to prepare to move to Guam.

During the decade, this basic organization underwent repeated redesignations of units and rearrangements of men and equipment within units. These occurred in response to changing operational requirements and deployments, to changes in naval air organization and nomenclature, and to growing specialization and sophistication in military aviation. They also often reflected the pressure of continued scanty appropriations, as Marine aviators tried to stretch limited manpower and equipment to meet their many responsibilities.

In 1922, the units at Quantico were redesignated collectively as the 1st Aviation Group and rearranged into three functionally specialized squadrons—one of observation planes, one of fighter planes, and one kite balloon squadron used for reconnaissance and artillery observation. Within the squadrons, flights now were retitled divisions to conform to Navy nomenclature of the time. Shortages of men and machines allowed each squadron to keep only one or at most two of its three divisions on active status. In 1924, with manpower made available by the Marine withdrawal from the Dominican Republic, the Marines formed Service Squadron 1, attached to the Aviation Group at Quantico. This non-flying unit contained truck drivers, riggers, mechanics, and other ground crew specialists. Like the flying squadrons, it had only one of its three divisions manned and active.

The terms "squadron" and "division" as used in the 1920s may be misleading to one familiar with today's aviation organization. Their meaning at this time, as well as the overall purpose of the organization adopted, was explained in 1926 by Major Brainard. Marine Corps aviation, he told students at Quantico:

> . . . is organized along Navy lines, with the division as the administrative unit and the squadron as the the tactical unit. In order to have an organization capable of large expansion in time of emergency, the peacetime squadrons are only one-third their war strength, i.e., one division active and two inactive. A division consists of 75 enlisted men and 10 commissioned officers. This gives the the nucleus around which to recruit the two inactive divisions, and the addition of

The F5–L flying boat, with its wingspan of 103 feet, was the largest and most modern patrol plane in the Marines' inventory in the early Twenties. (Nat Archives RG 127–G Photo 517539).

A Vought VE–7, a two-seater training version of this versatile and reliable, if slow, Marine aircraft of the early Twenties. (Nat Archives RG 127–G Photo 514919).

A single-seat VE–7F in 1922. Marines used these aircraft for a variety of missions. (Nat Archives RG 127–G Photo 515891).

Vought VE–9. In 1925, the 1st Aviation Group at Quantico had one of these, an improved version of the VE–7. (Nat Archives RG 127–G Photo 517532).

This O2B–1, in 1926, an improved DH–4B with a metal fuselage, carries on its side the name of its pilot, Major Charles A. Lutz, a successful Marine air racer. (Nat Archives RG 127–G Photo 525726).

a squadron commander and staff gives a war strength unit which should be fairly efficient and capable of shortly performing any task. A division consists of six planes active and three in reserve. Therefore, a full strength squadron has eighteen active planes and nine in reserve, and in addition two planes attached to squadron headquarters and one transport.[12]

In 1924, Marine aviation established itself on the West Coast when Observation Squadron 1 went by ship directly from the Dominican Republic to San Diego. The squadron became the nucleus of the 2d Aviation Group, created in 1925, which eventually consisted of one observation squadron, one fighting squadron, and one headquarters squadron. During the same period, the Marines were able to form two new squadrons at Quantico, one observation and one fighting. In August 1926, to complete the integration of aviation into the Marine expeditionary forces it was intended to support, Headquarters Marine Corps redesignated the 1st Aviation Group as Aircraft Squadrons, East Coast Expeditionary Forces, and the 2d Aviation Group as Aircraft Squadrons, West Coast Expeditionary Forces. This change put the squadrons under the supervision of the commanders of the respective Expeditionary Forces for purposes of training, administration, and operations.*

Besides these two principal Marine air groups, there remained throughout the 1920s Observation Squadron 2, as it now was designated, in Haiti and Scouting Squadron 1 on Guam, to which it had moved from Parris Island in 1921. The air units which deployed to China and Nicaragua later in the decade were drawn from the East and West Coast Expeditionary Force squadrons.

As the organization of the Marine air arm took shape during the 1920s, so did its concept of mission. In September 1926, in a lecture to student Marine officers, Major Brainard summed up the then-current doctrine. He defined three tactical missions for Marine aircraft: observation, which included artillery spotting and aerial photography; "light bombardment"; and "fighting

---

* A dispute arose between the Navy and the Marines over command of the West Coast squadrons. The question was whether they should be under the authority for administration, operations, and training of the commanding officer of the West Coast Expeditionary Force or of the commanding officer of the Naval Air Station at San Diego from which they were flying until the Marine field was completed. Finally, 2 September 1926, the Chief of Naval Operations put them under NAS San Diego until the Marine field was ready, when they would pass to West Coast Expeditionary Forces.

aviation," air-to-air combat to control the skies over Marine areas of operation. In Brainard's definition, this third category of operations also included low-altitude bombing and strafing of enemy ground troops. Turning to the broader reason for being of Marine aviation, Brainard continued:

> All our training and war plans are based on the idea that the Marine Corps will act as an advance base force to seize and hold an advance base from which the Navy can operate against the enemy. . . . In any war with a major force our fleet is going to be fully occupied and the advance base force will have to . . . use its own aviation for its information, protection from attack from the air and assistance in holding the base after seizure. I do not believe that the commander-in-chief is going to detach any first line carrier for this duty and for that reason Marine Corps Aviation is of paramount importance in the force. It also seems self-evident that there would be far better cooperation and results if the Marine force had Marine aviators rather than some Naval unit temporarily attached.

Brainard went on to urge the student officers to familiarize themselves with the techniques, problems, and potentialities of aviation, and he articulated the emerging vision of air and ground as a close-knit team. "To obtain maximum results, aviation and the troops with which it operates should be closely associated and know each other, as well as have a thorough knowledge of each other's work." Finally, in the light of controversies then raging in other services, he disavowed any aspirations to independence on the part of Marine aviation, declaring that "Marine Aviation is not being developed as a separate branch of the service that considers itself too good to do anything else. Unlike the Army Air Service, we do not aspire or want to be separated from the line or to be considered as anything but regular Marines."[13]

Men and Machines, 1920–1929

Throughout the decade, the authorized manpower of Marine aviation remained constant: 100 flying officers and 1,020 enlisted men. The actual number in service often fell much below these figures. In 1921, for instance, only 59 officers and 824 men were available for air duty; in 1923, the number dropped to 46 officers and 756 enlisted. Then it increased slowly and gradually during the rest of the decade.[14]

The principal source of officer-pilots remained the regular Marine officer corps. One function of the Aviation Section at Headquarters was to

Major Edwin H. (Chief) Brainard, head of Marine Aviation from 1925–29, beside his personal F6C–4 in 1927. (Nat Archives RG 127–G Photo 514772).

A rare photo of the first model of the Curtiss Hawk fighter, the F6C–1. There were only nine F6C–1s built. (Marine Corps Photo 527944).

enlist potential aviators from among the officer candidates and newly commissioned second lieutenants. Louis Woods remembered: "We did a lot of work trying to get pilots. We wrote letters. We looked over all the records and those we thought had the age and the background for aviation we tried to get. . . ." [15] In 1929, with accidents creating vacancies in pilot ranks and operations in Nicaragua demanding more men, Colonel Turner sent two of his pilots with a new Ford tri-motor transport to the officer candidate school, then located at Philadelphia, to sell the incoming cadets on aviation. For about a month, the two officers held informal ground training sessions every morning and then in the afternoons took planeloads of the students up for demonstration flights. The effort resulted in an increased number of applicants for aviation duty. [16]

Even if all 100 authorized officer billets could be filled, however, this number still fell short of the number of aviators the Marines needed to meet operational commitments. The Aviation Section could not enlarge the authorized number of flying officers because the Naval Appropriation Act of 4 June 1920 had fixed the commissioned strength of the entire Corps at 1,093, and the ground forces could spare no additional officers. Accordingly, Marine aviation attempted to remedy the pilot shortage by recruiting and training enlisted aviators. This was possible because the Appropriation Act limited only the officer strength—not the pilot strength. Through this means, the Marine Corps slowly increased its force of pilots. In a wartime expansion, the enlisted men so qualified could be commissioned and assume command positions commensurate with their experience and training.

Another potential source of men for wartime expansion was the Marine Aviation Reserve. After World War 1, when most Marine fliers had held reserve commissions, the reserve remained inactive from 1918 to 1928. In 1928, five Reserve aviators, most of them World War 1 veterans, were recalled to active duty. After brief refresher training courses they reported to Naval Reserve Aviation bases to organize new Marine Reserve Aviation units. On 1 July 1928, First Lieutenant Livingston B. Stedman formed a unit at Sand Point, Washington, while First Lieutenant Benjamin Reisweber organized a second one at Rockaway Beach on Long Island. Later in the month, two more units were organized, one at Squantum, Massachusetts, and the other at Great Lakes, Illinois.

Commanding officers of the reserve units selected applicants, mainly college students, for pilot training and commissions as second lieutenants, USMCR. Enlisted reservists were assigned to training as mechanics and ground crewmen. Officer candidates who met Marine mental and physical qualifications went on active duty at once for flight instruction, taking short basic courses at San Diego and Pensacola. They then

The VE–7F was a single-seat version of the VE–7 trainer used by the Marine Corps. This one, made by the Naval Aircraft Factory, was assigned to VO–1M in Santo Domingo, Dominican Republic, in 1922.

Major Alfred A. Cunningham at the controls of a DH–4B over Santo Domingo in 1922. (Photo from the Goodyear Kirkham collection).

received their commissions and later returned for one year active tours and advanced training with either the East or West Coast Expeditionary Forces.

James E. Webb, the future head of the National Aeronautics and Space Administration, who entered the reserve in the summer of 1930, recalled of the program:

> In those days they would give you one year to get your wings and then one year of active duty, and then you would be pushed out. The contract—they made it clear in the beginning they would not keep you beyond one year after graduation, because they wanted to train more. The whole purpose was to get more pilots available, then you could fly in the Reserve units on weekends to keep up your proficiency and be available to the service.

In fact, Marine reserve pilots on inactive status had to report twice a month at their own expense to the nearest Naval Reserve base to retain their flying proficiency.

In spite of inadequate funding, the Marine Aviation Reserve steadily expanded during the late 1920s. By the end of the decade, it had established solid foundations and was beginning to build toward its important role in World War II and subsequent national emergencies.[17]

New Marine pilots, both regular and reserve, received basic flight training at Navy fields, and regulars then usually took the Army pursuit course at Ellington Field, Texas, or Selfridge Field, Michigan. Both regulars and reserves received advanced instruction at Marine bases such as Quantico. This instruction included an increasing amount of practice "in working out tactical problems with troops on the ground." Enlisted Marines went through the mechanics' course at Great Lakes Naval Training Center, Illinois, and studied aerology, radio, and other technical subjects at the appropriate Navy schools. Throughout the decade, a few Marine officers and men continued to receive lighter-than-air flight training.[18]

Along with the training program went an increasing emphasis on air safety. Although aircraft performance and reliability had improved considerably by prewar standards, accidents continued to be frequent, almost expected as professional hazards of aviation. The number of fatalities, listed with grim regularity in every annual report on Marine aviation, caused concern at all levels of command. Certainly one cause of the many smashups was the inadequacies of early Marine landing fields. The words of veterans recalling those days bring the dangers vividly to life. At Quantico in 1922: "They only had one field there that was on the east side of the tracks. We came in over about a 30 foot bank, the bank of the river, and the other side had high tension wires. The wind was always cross wind—it was always from the north-west."[19] At Parris Island:

> Our field was the old, old rifle range, which was nothing but a pocket handkerchief. . . . Those of us who went to Parris Island, having to operate out of that extremely tiny field, got more experience than anyone in the world. We could go into anything, any size field, that no one in the world could go into, because we'd had to, or get killed. That was the best training that ever was.[20]

Safety programs were instituted throughout Marine aviation to reduce the accident rate. While unsophisticated by today's standards, these efforts were on a scale commensurate with the number of aviators and the number of hours flown. They included a review of pilot indoctrination practices, the establishment of some standard aircraft flying procedures, and an attempt to reduce the number of unscheduled individual aerial stunt performances. At least one Marine aviator was successful in his safety program. Captain Harold D. Campbell received the Schiff Trophy, the annual air safety award, for having flown 839.50 accident-free hours during the fiscal year 1925–1926. For presentation of the trophy by the President of the United States, Captain Campbell flew from San Diego to Washington, D.C., a flight which was still an accomplishment in itself in 1926.[21]

Marine aviators trained and flew operations in a variety of aircraft most of which they obtained from the Army and Navy, sometimes as cast-offs when the other services adopted newer models. The Marine squadrons sent to Haiti and the Dominican Republic in 1919 deployed with Curtiss Jennies and HS–2L seaplanes. That same year, the Marine Corps received from the Army an assortment of surplus aircraft, including 15 DH–4Bs, half-a-dozen Fokker D–7s, and 11 Thomas-Morse Scouts (now numbered MB–3s).

An improved version of the sturdy mainstay of the 1st Aviation Force, the DH–4B performed a variety of missions for the Marines throughout the decade, serving as dive bomber, observation plane, light transport, and even—remodelled to carry one stretcher patient at a time—as a hospital plane. Frequently rebuilt by several aircraft firms, it emerged and re-emerged in numerous versions and guises. The Boeing O2B–1 observation plane, for example, was a DH–4B rebuilt with a metal-framed fuselage.

The Marines made little use of the Fokker D–7s. They left most of them in their crates except for two which Second Lieutenant Sanderson and "Tex" Rogers, then a first lieutenant, assembled on their own as personal aircraft. While Rogers called his D–7 a "dream plane," it evidently took some getting used to because after Sanderson and Rogers left for the Caribbean other aviators promptly crashed and were killed in both machines. The Thomas-Morse Scouts, which arrived from the Army in very poor mechanical condition, were the first true fighters to come into the Marines' hands and went to equip Flight F of 3d Squadron at Quantico, the Marines' first specialized fighter unit. Powered by a French rotary engine which developed excessive torque and rudder pressure and hence according to pilots "fast, tricky and as tiring as hell to fly," the Thomas-Morse gave the Marines an early chance to practice fighter tactics.

In 1921, Marine units began receiving the new Vought-designed VE–7 and VE–7SF. Essentially an improved version of the Jenny, this aircraft had been designed during the war as a trainer but also proved usable as a fighter, scout, and observation craft. The Navy purchased large numbers of this machine from Vought and arranged for its production by the Naval Aircraft Factory. While slow, the VE–7 and its scout-fighter model, the VE–7SF, were reliable and maneuverable. The Marines used them for a variety of missions in single-seat, dual-seat, and float plane versions.[22]

In the early 1920s, the Marines received a new flying boat and one of the largest bombing planes of the time. The flying boat, the F5–L built by the Naval Aircraft Factory, was the largest and most modern patrol plane the Marines yet had possessed. Powered by two modified high-compression Liberty engines, it had a wingspread of over 103 feet and carried a

The Curtiss F8C-1 and subsequent F8C models served the Marines as fighters, dive bombers, and observation planes. This photograph was taken about 1923. (Marine Corps Photo 529613).

A Marine F-5-L flying boat on Guam, about 1923. (Marine Corps Photo 514933).

Vought O2U–1 "Corsair" in 1928. This aircraft took over many of the missions performed by the O2B–1. (Marine Corps Photo 529952).

A "Corsair" of VO–7M, the squadron which operated against Sandino in Nicaragua. These maneuverable, reliable aircraft more than proved their worth in the Nicaraguan campaign in 1929. (Marine Corps Photo 526365).

gunner, a radio operator, and a crew chief besides its pilot and co-pilot. It was equipped with the latest model radio compass and its engine ignition system had been modified to prevent interference with this sensitive navigational equipment. An extremely heavy plane, the F5–L was very difficult to get out of the water on takeoff. Once in the air, pilots found it hard to maneuver.

In 1922–1923, the aviation group at Quantico received six Martin heavy bombers. Powered by twin Liberty engines, each of these machines had a wing span of 71 feet, five inches, and a length of 46 feet, five inches. Each had four landing wheels arranged side by side to support its weight of 12,078 pounds and carried a three-man crew. A pilot who flew the Martin bomber, which the Marines used mainly as a transport, recalled it as "quite slow, 80 knots I think," but "an easy plane to fly. It was heavy, but it actually controlled very nicely." [23] The Marines also used the big planes to carry parachute jumpers.

Marine airplane strength in the mid-1920s was far from impressive. In 1925, for example, the 1st Aviation Group had 25 operational aircraft, including nine DH–4B1s, six VE–7s of various types, one VE–9 (an improved VE–7), six Martin heavy bombers, two trainers, and one O2B–1. Observation Squadron 1 at San Diego had six operational DH–4B1s and one Jenny. Observation Squadron 2 at Port au Prince had two DH–4Bs and three DH–4B1s, while Scouting Squadron 1 on Guam had three HS–2Ls. [24]

In 1925 and 1926, after investigations of the Army and Navy air services and the aircraft industry by congressional committees and a special Presidential panel, the Morrow Board, the services adopted and Congress funded a new aircraft construction and development program. [25] As a result, Marine aviation along with the other air services received during the late 1920s a succession of new aircraft superior in design and performance to anything Marines had flown previously. These aircraft included a new generation of fighters—the Boeing FB series and the Curtiss F6C series. Of mixed metal and wood-and-fabric construction, these biplanes had top speeds of about 150 miles per hour and service ceilings of about 21,000 feet. They could climb about 5,000 feet in two and a half to three minutes. Capable of making almost vertical dives, planes of the Curtiss F6C series gave naval and Marine aviators for the first time a craft which could execute effectively their new tactic of dive bombing. A Marine who flew the F6C described aircraft of that type as:

. . . beautiful planes to fly, they were very maneuverable, easy to maneuver, and quite safe, too. I mean you could hardly ever get yourself into difficulty with them, within reason. They'd spin like a top. The easiest thing to get to spin and the easiest to get out . . . . All you had to do was let go of the controls and it came out itself. It was a well trained airplane! [26]

Hardly had the FBs and F6Cs come into service when a new engine revolutionized the design of Marine and naval aircraft. This was the air-cooled radial engine, perfected in the mid-1920s and put into production initially by the Wright Aeronautical Corporation and Pratt and Whitney. Lacking the complicated and trouble-prone liquid cooling systems of earlier power plants, the radial engine was simpler in design, easier to maintain, and could deliver more horsepower for the same size and weight.

Curtiss quickly adapted its F6C–4 "Hawk" for the new engine, and this plane served as the Marines' standard fighter throughout the late 1920s, proving maneuverable and easy to fly. The Marines also acquired a new radial-engine observation plane, the Vought O2U "Corsair," which took over many of the missions hitherto performed by variants of the DH–4. A pilot who flew the Corsair in Nicaragua called it "An outstanding combat plane: very light, had a lot of power in it. . . . When we got that down in Nicaragua we were very happy to get it because of outstanding performance, ease in handling, quick takeoffs and landings."

Other radial engine machines received by the Marines included the Curtiss F7C "Sea Hawk" and the F8C–4 "Hell Diver" which functioned at different times as a fighter, dive bomber, and observation plane. In the form of the new Atlantic-Fokker tri-motor monoplane, supplemented in 1929 by the Ford tri-motor ("Tin Goose"), the Marines developed an air transport capability. [27]

Maneuvers, Long-Distance Flights, and Air Races

With aircraft that gradually improved in performance and reliability, Marine aviators during the 1920s conducted a wide variety of operations. In this publicity-conscious decade of peace and relative prosperity, they found it necessary to combine serious military exercises with headline-hunting spectaculars.

At home stations and overseas, Marine squadrons conducted scheduled training exercises which included annual bombing and machine gun practices. The 1st Aviation Group at Quan-

Having a pre-flight cigarette are, from left: Admiral Moffett, Director of Naval Aviation; Lieutenant Colonel Thomas C. Turner; Lieutenant Bradley; Sergeant Rucker; and Lieutenant Lawson H. M. Sanderson. The Marines flew two planes from coast-to-coast in 1921. (Photo from the Goodyear Kirkham collection).

Goodyear W. Kirkham at the controls of his Thomas Morse Scout (MB–3). (Photo from the Goodyear Kirkham collection).

A Curtis F6C-4 Hawk in 1928. The exposed cylinders of the air-cooled radial engine contrast with the earlier liquid-cooled models. (Marine Corps Photo 515095).

Curtis Marine Trophy winner Major Charles A. Lutz with a Curtiss "Hawk" F6C-3 in 1927. The triangular pennant on the aircraft's side denotes excellence in gunnery. (Marine Corps Photo 530813).

A Boeing FB–1 fighter on the Marine landing strip at Tientsin in 1927. Aircraft such as this supported Butler's brigade in China. (Marine Corps Photo 514929).

One of the O2B–1s of the observation squadron in China. When first formed, this unit was numbered VO–5M, but by late 1927 had been redesignated VO–10M, as the side markings of this aircraft indicate. (Marine Corps Photo 514928).

Lieutenant Christian F. Schilt, beside his Navy racer after placing second in the Schneider Cup race of 1926. (Marine Corps Photo 524432).

An OL–9 Loening amphibian circa 1933. Marines made extensive use of these aircraft in Nicaragua and elsewhere. (Marine Corps Photo 515519).

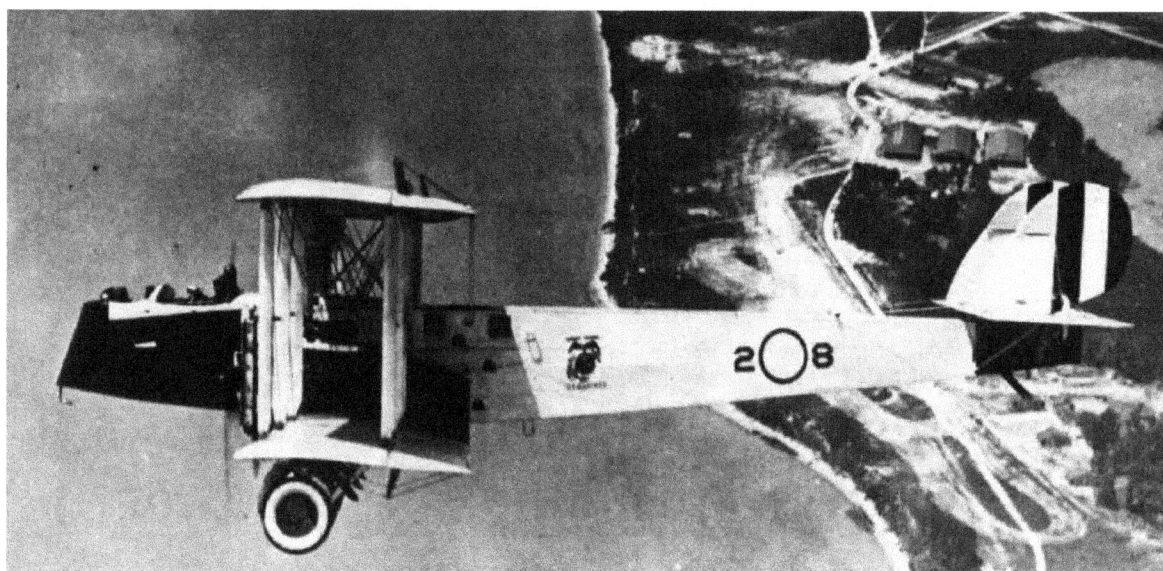

A Martin MBT flying over Quantico around 1926. Of the 10 MBT built, at least six were placed with the Marines. (National Archives Photo 80-G-4144).

tico practiced artillery observation, tested methods of air-ground communication, and engaged extensively in aerial photography and mapping. In the summer of 1921, DH–4s of this group took part in the experimental bombings of captured German battleships by which Brigadier General "Billy" Mitchell, USA, dramatized the ability of aircraft to sink stationary and undefended capital ships.

The 1st Aviation Group annually participated with the ground troops at Quantico in the large scale maneuvers instituted by Brigadier General Smedley D. Butler. In the most spectacular of these in 1922, aircraft of the group including three of the big Martin bombers participated in a practice march of 4,000 Marines from Quantico to Gettysburg, Pennsylvania, where the troops re-enacted Pickett's Charge and then conducted a modern battle exercise on the same ground. Logging a total of 500 hours and 40,000 miles in the air, the planes carried passengers and freight and maintained radio contact with the column, executing attack missions assigned by the ground commanders. Following the trend of the times, this operation combined real training with a publicity spectacle.[28]

While Marine aviation began to develop a coherent training program during the decade, much of the activity of the pilots seemed undisciplined and haphazard by later professional standards. Some post commanders, including Brigadier General Butler, took an interest in aviation and sought to improve the training

and expand the capabilities of the squadrons under their charge. Others, as a Marine veteran recalled, "didn't know what to do with us." Individual pilots enjoyed great freedom of action. As one of them described it:

> In the 20's, there were no restrictions. You could go anywhere you wanted to go, any time. You had a cross country packet where you were authorized to buy gasoline. The cross country packet gave you shipping orders if you cracked up, to ship your plane back, if you went anywhere at all. . . . If we wanted to go someplace we'd just go, and never asked anybody's permission or a blessed thing. Just go. . . .[29]

To demonstrate professional skill, test new equipment and techniques, and capture for their service a share of the decade's stunt-filled headlines, Marine aviators made long-distance flights and participated in air races. Lieutenant Colonel Turner, soon after taking over as head of the aviation section, personally joined in this effort. In April 1921, to demonstrate the practicality of long-distance aircraft deployments, he led a flight of two DH–4s from Washington, D.C., to Santo Domingo, establishing a record for the longest unguarded flight over land and water made up to that time by American Navy or Marine personnel.[30]

Two years later, Marine pilots led by Major Geiger ferried three of the giant Martin bombers across the United States from San Diego to Quantico. Travelling from California across Arizona, New Mexico, and Texas, then north up the

Mississippi Valley and eastward across Illinois, Indiana, Ohio, and West Virginia, the Marine aircrews took 11 days to make the trip, counting in frequent stops for fuel and repairs. They navigated with Rand-McNally maps and if they became lost, they flew low over railroad stations to read the names of towns off the signs. After a ceremonial welcome at Washington, D.C., they finally landed at Quantico on 30 April 1923.[31]

In 1923, Marine aviators made their most impressive long-distance flight, when two DH–4s flew from Haiti to Saint Louis, Missouri, to attend the National Aircraft Races for that year. Marine Corps Headquarters authorized the flight, which was an attempt to capture for the Marines the distinction of having flown a plane the longest distance to attend the race. Three weeks before the planned departure date of 20 September, two new DH–4s and two Liberty engines were shipped in crates to the Marines' intended starting point in Haiti. The aircraft and engines reached Port au Prince on 15 September, and when they were uncrated and checked, the Marine mechanics found that the engines were defective. Working in shifts for four days, 24 hours a day, the ground crewmen rebuilt the engines, assembled the air frames, and installed dual controls in each aircraft.

At noon on 19 September, the two DH–4s took off from Santo Domingo City. First Lieutenant Ford O. ("Tex") Rogers piloted one of the machines, with First Sergeant Benjamin F. Belcher as his observer. Second Lieutenant Horace ("Hoke") Dutton Palmer flew the second, with Sergeant Peter P. Toluscisk, as observer. All of these Marines came from the air squadrons then stationed in Haiti and the Dominican Republic.

Flying a route that led them past Port au Prince, Guantanamo, and Havana, they crossed to the United States mainland at Key West and then flew on to Washington, D.C., where they landed on 23 September. After a day's rest, they headed for St. Louis, arriving on schedule five days after they left Washington. From St. Louis, they extended their flight on across the continent to San Francisco, at the invitation of American Legion headquarters in that city. Then they flew back from San Francisco to Washington again, where they touched down on 2 November and stopped over to make repairs to the aircraft and install new engines.

On 30 November, they left Washington for their home base in Santo Domingo. The return trip proved uneventful with the exception of a hard landing at Miami by Lieutenant Palmer,

From September to November 1923, 1st Lieutenant Ford O. (Tex) Rogers flew his DH–4 10,953 miles in 127 hours of air time to demonstrate the skill of Marine pilots and aviation mechanics. (Photo from the Goodyear Kirkham collection).

which resulted in a bent landing gear axle. Marine mechanics repaired it by using a car jack, and the flight departed on the final overwater leg to Santo Domingo, where they landed triumphantly on 9 December.

In some two and one-half months, these Marine aviators had travelled 10,953 miles in about 127 hours of actual flying time. They had fallen just short of the 13,500 miles flown by British aviators flying from England to Australia. The exploit dramatically demonstrated the skill of Marine pilots and also that of Marine aviation mechanics, who had put the two DH–4s in such excellent condition that they flew for 110 hours without so much as a spark plug change.[32]

Air races became an American institution in

A later model of the Boeing fighter, un FB-3 of VF-1M at the Philadelphia Air Races of 1926. (Nat Archives Photo 80-C-452008).

the 1920s and retained their popularity until the Second World War. Every large state and county fair with any pretensions to success had to have an air race or show. Major national races, such as the Pulitzer, provided the Services with a method of determining which combinations of air frame design and engine produced the best performance and they kept the American public aware of aviation—in particular military aviation—at a time when there was no opportunity to demonstrate its applications in a world-wide conflict.

As was so often true of Marine aviation, much Marine participation in these races occurred under Navy colors, with individual Marine pilots flying one or more of the Navy planes entered. In 1926, for example, Lieutenant Christian F. Schilt, USMC, soon to win the Medal of Honor in combat in Nicaragua, flew one of the three Navy planes entered in the prestigious Schneider Cup race and came in second. Such triumphs, while gratifying to Marines, often were not fully recognized by spectators who failed to realize the Navy aircraft they were watching had Marine pilots. Thus Marines took more satisfaction in the exploit of Major Charles A. Lutz, who captured first place in the Curtiss Marine Trophy Race at Anacostia Naval Air Station on 19 May 1928 while flying a Marine Curtiss Hawk. Lutz achieved an average speed of 157 miles per

hour for five laps over a closed course of 20 miles. Further sweetening the victory, another Marine, Captain Harold D. Major, also flying a Hawk, took third place.[33]

Operations in Haiti, the Dominican Republic, and Guam

While long-distance flights and air races caught the attention of the newspapers and the public, Marine aviators were gaining operational experience that in the long run was far more significant for the development of their service. During the 1920s, Marine brigades were sent to the Dominican Republic, Haiti, China, and Nicaragua to protect United States national interests. Marine air units, often added to these expeditions as an afterthought and lacking any clear advance directive as to their role, for the first time had a chance to demonstrate their ability to assist the operations of Marines on the ground. As one aviator summed it up, "We were there and they used us, and they used us to their advantage, and consequently we became a useful and integral part of the Marine Corps." [34]

Overseas operations began in 1919 when air units were sent to support existing Marine commitments in Haiti, occupied in 1915, and the Dominican Republic, where Marines had landed

An Atlantic-Fokker TA–2 at Managua, Nicaragua, in 1929. These tri-motor transports greatly increased the freight and passenger carrying capacity of Marine aviation. (Marine Corps Photo 528145).

The "Tin Goose" or Ford RR–2 trimotor transport, further enlarged Marine airlift capacity in Nicaragua in 1929. Its all-metal construction made it easier to maintain than the Fokker. (Marine Corps Photo A402978).

One of two O2U–1 "Corsairs" flown by Lieutenant Schilt at Quilali in 1929. The oversize DH-4B wheels dominate the undercarriage. (Marine Corps Photo 529590).

in 1916. The six Jennies of 1st Air Squadron, commanded by Captain Walter E. McCaughtry, began operations at San Pedro de Macoris, Dominican Republic, in February 1919, while the six Jennies and six HS–2Ls of the 4th squadron*

---

* This force was designated 1st Division, Flight E, until the reorganization of 1920.

under Captain Harvey B. Mims took station at Port au Prince, Haiti, on 31 March.

The 1st Squadron operated in the Dominican Republic until 1924, when it withdrew with the rest of the Marine contingent. The squadron in Haiti remained with the Marines in that country until final American evacuation in 1934. In both countries, Marine aviators assisted their com-

A Fokker transport prepares to drop supplies to a patrol in Nicaragua in 1929. (Marine Corps Photo 514940).

A Curtiss HS–2L of Marine Scouting Squadron 1 (VS–1M) on Guam in May 1926. (Marine Corps Photo 530811).

Three of the giant Martin bombers lined up on the field at Quantico in 1925. (Nat Archives RG 127–G Photo 514939).

A Boeing FB–1 of Marine Fighting Squadron 6 (VF–6M) at San Diego. This was an early model of the new generation of aircraft the Marines began receiving in the late Twenties. (Nat Archives RG 127–G Photo 530238).

rades on the ground in drawn-out, tedious guerrilla warfare against indigenous irregulars, called "Cacos" in Haiti and "Bandits" in Santo Domingo.

Aircraft on a number of occasions took part in active combat, bombing and strafing bandit groups or guiding ground patrols to contact. However, the limited armament and maneuverability of the planes and the lack of rapid, reliable air-ground communications rendered Marine aviation less than decisive as an anti-bandit weapon. In both Haiti and Santo Domingo, the air squadrons proved most useful in indirect support roles, carrying mail and passengers to remote posts, reconnoitering and mapping, and sometimes transporting supplies or evacuating wounded men. The ability of aviation to enhance the mobility of forces operating in largely roadless terrain began to become apparent to Marines in these campaigns.[35]

During the operations in Haiti, Marines began practicing a tactic fundamental to the carrying out of their close support mission. That tactic was dive bombing. During the summer of 1919, Lieutenant Lawson H. M. Sanderson of 4th Squadron, then stationed in Haiti, decided that he and his fellow pilots needed a more accurate method of delivering bombs against the enemy "Cacos." In experimental exercises, Sanderson abandoned the hitherto standard procedure of allowing his observer to release the bomb from horizontal flight while aiming with a crude sight protruding from the rear cockpit. Instead, he entered what was then considered a steep dive of 45 degrees, pointed the nose of his aircraft at the target, and released the bomb from the pilot's position at an altitude of about 250 feet.* He found that this method made his bombing much more accurate, and other members of his squadron soon adopted it. By late 1920, Marines at Quantico were using it also.[36]

While Sanderson introduced dive bombing to Marine aviators on the Atlantic Coast around 1920, it reached West Coast Marines from the Army. In May 1923, while taking an advanced course of instruction at Kelly Field, Texas, Major Ross Rowell, USMC, observed and participated in dive bombing exercises directed by Major Lewis H. Brereton, USA. Rowell, who claimed that this was the first time he had seen dive bombing, was impressed with its accuracy and "I immediately visualized the certain naval employment of such tactics where accuracy against small moving targets is paramount. Also it seemed to me that it would be an excellent form of tactics for use in guerrilla warfare."

When he took command of Observation Squadron 1 (VO–1M) at San Diego in 1924, Rowell trained his pilots in dive bombing and obtained Army-type, wing-mounted bomb racks for their DH–4Bs.** His squadron put on dive bombing demonstrations at airport openings and air shows all up and down the West Coast. Eventually in Nicaragua they would have the chance to use their skill in combat.[37]

While Sanderson, Rowell, and others experimented with new tactics, Marine aviation in 1921 began its historic role in the Pacific when Flight L, organized at Parris Island, went by ship to Sumay, Guam. Since no air facility then existed on Guam, the unit's first mission was to build an airfield and seaplane base as part of a Navy plan (aborted by the Naval Disarmament Conference of 1921–1922) to build up the island's defenses. To this end, the flight embarked with every spare piece of air station equipment the Navy and Marine Corps could gather from the East Coast. For aircraft, the flight received N–9s and HS–2Ls, along with the giant F–5L. Later the unit acquired VE–7s and Loening amphibians. After completing its base on Guam, the unit settled down to routine training and the collection of meteorological data, continuing both activities until it was withdrawn from Guam in 1931. The weather information gathered by these Marine aviators, along with the presence of the

---

* By modern standards, what Sanderson was doing would be called "glide bombing," as a true, steep, powered dive was impossible in the planes of that day. At the time, however, they called it dive bombing and with sturdier machines like the Curtiss F6C series began to approximate the modern tactic. Lieutenant Sanderson never claimed to be the inventor of dive bombing, although probably he was the first Marine to use the tactic. Apparently, dive (or glide) bombing evolved in a number of air services during World War I. Both Allied and German pilots are reported to have used it in combat, and U.S. Army fliers at Ellington Field, Texas, practiced it during 1917–1918, dropping their bombs from wing racks controlled by wires leading to the pilot's cockpit.

** Marine aviators during the 1920s used any scout or observation plane for dive bombing, including Jennies and DH–4Bs and later Curtiss Hawks and Helldivers. Biplanes could dive bomb without wing flaps or diving brakes because their "built-in headwind" of struts, wires, fixed landing gear, etc. kept their speed under 400 miles per hour even in a wide-open vertical dive.

air facilities that they built, contributed much to the development of trans-Pacific aviation.[38]*

China and Nicaragua

New overseas commitments developed in 1927, when the outbreak of civil wars in China and Nicaragua threatened American lives and interests in those countries and resulted in the dispatch of Marines. As in Haiti and Santo Domingo, Marine aviation accompanied the expeditions. To support Brigadier General Smedley D. Butler and his 3d Brigade in China, Fighting Squadron 3 (VF–3M) sailed from San Diego for Shanghai on 17 April 1927 with 9 officers, 48 enlisted men, and 8 FB–1s. It was reinforced by a new observation squadron (VO–5M) which was organized in China with aircraft (six O2B–1s) sent from San Diego and four officers and 94 men from the unit on Guam. These deployments made the Marine brigade, when it moved up to Tientsin, the center of trouble, the only foreign contingent in the area with its own aviation.

---

* Cmdr G. C. Westervelt (C.C.) U.S. Navy and H. B. Sanford, Aeronautical Engineer, "Possibilities of a Trans-Pacific Flight," *United States Naval Institute Proceedings,* v. XLVI, No. 5 (May 1920), pp. 675–712. This article proved academically the possibility of making a trans-Pacific flight in a Navy NC-type flying boat, a type which recently had flown the Atlantic. The article presented in detail flight plans for several routes depending on the wind conditions of the season. Guam played a vital role in all the plans.

Commanded initially by Major Francis T. ("Cocky") Evans and then by Lieutenant Colonel Turner, the Marine squadrons stayed in China for a year and a half. They operated from a pasture levelled into a flying field by coolie labor about 35 miles from Tientsin. Isolated from the rest of the Marine brigade and with columns of troops from the rival Chinese armies frequently marching past them, the Marines formed their own base guard detachment and mounted machine guns on their hangars and barracks. No combat occurred for these Marines, however, either in the air or on the ground. The squadrons flew 3,818 sorties in support of the Marine brigade's peace-keeping mission. They spent most of their time in observation and photographic reconnaissance, tracking for General Butler the movements of the Chinese forces. They also carried mail and passengers.[39] The airmen's professional competence received high praise from Butler, who said in a message to Turner:

> Our aircraft squadrons . . . have not been surpassed in their efficiency. Not only did they never fail immediately and successfully to respond to all calls, but they maintained themselves in the open for nearly eighteen months and at all times in readiness. . . . Their performance at all times was brilliant. . . . There has not been one fatality or serious injury.[40]

In 1929, as conditions quieted down in China, these units returned to their former stations at San Diego and Guam.

A lineup of Boeing FB–1s of Marine Fighting Squadron 2 at Quantico in 1926. (Marine Corps Photo 515863).

The Marine Corps received five Curtiss F7C–1s in January 1929. This was the personal plane of Captain James T. Moore, CO Air Service, East Coast Expeditionary Force, Quantico. (Marine Corps Photo 517619).

In Nicaragua, Marine aviation became involved in a small-scale, but drawn-out and difficult guerrilla war during which for the first time Marine fliers regularly gave something resembling close air support to troops engaged in ground combat. In 1927, the outbreak of civil war in Nicaragua* led to Marine intervention. Under the "Stimson Agreement," named after American negotiator Henry L. Stimson, leaders to both warring Nicaraguan factions agreed to disarmament of their troops and to an American-supervised national election. Stability collapsed again when Augusto C. Sandino, a general of the Liberal faction, denounced the Stimson Agreement and declared war on both the Marines and the Nicaraguan government. There followed years of sporadic bush fighting which continued until the early 1930s.

Two Marine air squadrons entered Nicaragua with the initial intervention force. On 18 February 1927, Observation Squadron 1 (VO–1M), with 8 officers, 81 enlisted men, and 6 DH–4Bs, embarked at San Diego for the Nicaraguan port of Corinto. Unloading from their transports there, they travelled by train to Managua with their aircraft, with the wings removed, carried on flatcars. At Managua, the squadron established itself in the ball park on the edge of the city, where the Marines remained for four months and from which they operated in co-operation with the Nicaraguan air force.* VO–4M from Quantico, with seven officers and 78 men equipped with six O2B–1s sailed on 21 May to reinforce VO–1M. Upon its arrival in Nicaragua, the two units were designated Aircraft

---

* Nicaragua had strategic importance for the United States because it contained within its borders an important alternate inter-oceanic canal route.

* The Nicaraguan air force consisted of two barnstorming pilots flying Laird Swallow aircraft which, according to Major Rowell, were "discards from the Checkered Cab Co., at San Francisco."

Squadrons, 2d Brigade, and placed under the command of Major Ross E. Rowell.

From February until May of 1927, aircraft of these two squadrons flew patrols over the neutral zone established and occupied by the Marines, and they conducted visual and photographic reconnaissance flights over the lines of the hostile Nicaraguan armies. During this period, under directions from Washington, the Marine airmen engaged in no combat beyond a couple of machine gun attacks on rebels who penetrated the neutral zone. In June, with order seemingly restored by the Stimson Agreement, most of VO–1M returned to San Diego. A few men and two of the Squadron's DH–4Bs remained with VO–4M, which was redesignated VO–7M on 1 July 1927. Major Rowell stayed in Nicaragua to command the reorganized squadron.[41]

On 16 July 1927, Sandino explosively demonstrated that hopes for stability were premature. At 0115 on that day, with an estimated force of 500 men, he attacked the town of Ocotal.[42] The garrison of 38 Marines and 49 Nicaraguan National Guardsmen rallied quickly and repulsed the first attack. Further unsuccessful rebel assaults followed until 0810, when Sandino made a truce offer that was refused by the defenders. The attack then resumed. The position of the Marines and guardsmen was precarious. Ocotal lay some 125 miles away from Managua, where most American forces were concentrated, and by ground transportation it would take a relief force 10 days to two weeks to cover that distance. The garrison had only limited stocks of water, food, and ammunition.

In this, the first major action of Sandino's war, Marine aviation intervened with dramatic and decisive effect. Around 1030 on the morning of 16 July, the routine daily reconnaissance patrol of two aircraft, piloted by Lieutenant Hayne D. Boyden, and Gunner Michael Wodarcyzk, arrived over Ocotal. Observing the situation from the air, the two aviators moved to aid the garrison. Boyden, who lacked radio contact with the ground, landed to obtain information from a villager. Wodarcyzk began strafing the bandits to protect Boyden. Boyden then took off for Managua to make his report while Wodarcyzk continued his strafing attacks around Ocotal for another 20 minutes.

As soon as he received Boyden's report, Major Rowell ordered his five available DH–4Bs and O2B–1s armed and fueled. He forwarded the report to the brigade commander, Brigadier General Logan Feland, and received in reply orders "to take such immediate steps as I deemed to be most effective in succoring the besieged Marines and Guardia." At 1230, Rowell and his flight took off from Managua. Each aircraft carried a full combat allowance of 600 rounds of ammunition for each of its machine guns but only a partial load of bombs due to the fact that the planes had to carry a heavy fuel load for the long flight.

The trip to Ocotal took about two hours because Rowell's formation had to fly around a line of thunder storms. Around 1435, they arrived over Ocotal. Rowell had trained all of his pilots in dive bombing and planned to use that mode of attack. Putting the flight into column formation, he led one circle of the town to locate enemy and friendly positions, then launched his assault. As Rowell later described the 45-minute action:

> I led off the attack and dived out of column from 1,500 feet, pulling out at about 600. Later we ended up by diving in from 1,000 and pulling out at 300. Since the enemy had not been subjected to any form of bombing attack, other than the dynamite charges thrown from the Laird-Swallows by the Nicaraguan Air Force, they had no fear of us. They exposed themselves in such a manner that we were able to inflict damage which was out of proportion to what they would have suffered had they taken cover.[43]

In their diving attacks, Rowell and his pilots fired their front machine guns on the way down and dropped fragmentation bombs when targets presented themselves. As they pulled out of their dives their observers strafed the Sandinistas with their rear cockpit guns. After the second pass by the planes, bandits began fleeing out of the town, along with stampeding horses. Reports on the number of casualties inflicted on Sandino's men are conflicting, but, as the commander of the ground defenders of Ocotal stated in his report, "The air attack was the deciding factor in our favor, for almost immediately the firing slackened and troops began to withdraw." [44] Thus ended what probably deserves to be called the first Marine air-ground combined action.

After Ocotal, Sandino usually did not mass his forces where aircraft could reach them. He maintained his hit-and-run war year after year while the Marines and the National Guard launched operation after operation against him. VO–7M, reinforced after February 1928 by the 2 officers, 59 enlisted men, and 6 O2B–1s of VO–6M from Quantico,[45] provided combat, reconnaissance, and logistical support for these efforts. The arrival late in 1927 of the first new Vought O2U "Corsairs" improved the squadrons' capabilities.

Flying one of the newly arrived Corsairs, Lieutenant Christian F. Schilt gave a courageous demonstration of the airmen's ability to aid hard-pressed infantry. On 30 December 1927, a patrol encountered a large Sandinista force near the village of Quilali. After a firefight in which the Marines took heavy casualties but drove off the bandits, the patrol took up defensive positions in the village. Reinforcements were sent from the nearby town of Telpaneca, but the relief column came under fire about five miles from Quilali. It took several air attacks and a patrol from Quilali to disperse the bandits and permit the two patrols to consolidate their defenses in the village. By this time, most of the commissioned and noncommissioned officers of both patrols had been killed or seriously wounded. In fact, a total of 18 wounded men needed immediate evacuation if they were to survive and if the patrols were to recover mobility. The acting commander of the beleaguered force, in a message relayed to headquarters in Managua, asked for air attacks to break up the bandit concentration surrounding him and recommended that "if humanly possible" a Corsair land at Quilali to take out the wounded.

In response to this message, Marine pilots dropped tools, supplies, and equipment to the defenders of Quilali, who cleared away the jungle and part of the village to create a rough, hole-pocked strip about 500 feet long. Lieutenant Schilt, in a Corsair fitted with over-sized wheels to negotiate the treacherous runway, made ten trips into the hastily prepared landing field on 6, 7, and 8 January 1928. On one of his first flights he brought in a new commanding officer along with badly needed medical supplies. In all, he flew in about 1,400 pounds of stores and evacuated the 18 seriously wounded. For this aeronautical accomplishment and display of pure courage, Lieutenant Schilt received the Medal of Honor.[46]

After Quilali, Marine aviation took part in many operations against Sandino. In January 1928, aerial reconnaissance and a preliminary bombing and strafing attack prepared the way for a major Marine-National Guard assault on Sandino's supposed mountain-top stronghold of El Chipote. The attack inflicted bandit casualties, but once again the elusive Sandino and most of his men escaped the net. Later in the same year, Marine air strikes severely punished a large enemy force at Murra, near Ocotal. Over the next four years, dive bombing and strafing

attacks in support of ground troops, sometimes directed from the ground by colored panels or other signalling devices, became a routine feature of operations. Neither side could claim decisive victories in this bush war, but the continuous pressure and aggressive tactics of the Marines began to show substantial results as early as the summer of 1928. From May to July of that year, more than 1,000 guerrillas surrendered to the Nicaraguan government under the promise of amnesty. Sandino and his hard core followers remained in the field, however, until 1931.[47]

Besides assisting Marines in combat, the air arm in Nicaragua enlarged its air transport role, using the newly acquired Atlantic-Fokker tri-motors. The first of these machines landed at Managua on 4 December 1927, ferried down from the United States by Major Brainard. During its first six weeks of operation, this transport carried 27,000 pounds of freight and 204 passengers, most of them on the long flight between Managua and Ocotal. The tri-motor could make this trip, which took ox-carts or mule trains 10 days to three weeks, in one hour and 40 minutes. Under Nicaraguan conditions, it could carry 2,000 pounds of cargo or eight fully equipped Marines per flight. So useful did this plane prove that two additional ones soon were put into service in Nicaragua. They were supplemented later by all-metal Ford tri-motors, which required less maintenance in the tropical climate than did the Fokkers with their canvas and wood wings. Able to fly six tons of supplies per day from Managua to Ocotal, Major Rowell set up an advanced air base at the latter city, which was closer to the bandit regions than was Managua.[48]

As fighting slackened off in Nicaragua after 1928, the Marine squadrons concentrated on observation, medical evacuation, and logistical support missions. They established a scheduled mail and passenger service to assist both the American forces and the Nicaraguan government. They also did extensive aerial mapping and photography.

A Decade of Achievement

Marine aviation in 1929 could look back upon a decade of significant progress and achievement. Although hampered by low budgets and often forced to operate with outmoded or cast-off equipment, Marine aviators during these years perfected a stable organization. They formulated a mission and began to train themselves to

A squadron leader's FB–1 of VF–1M in flight over Quantico in 1928. (Marine Corps Photo 530238).

perform it. In Santo Domingo, Haiti, China, and Nicaragua, they adopted and refined new tactics, such as dive bombing, for carrying out their mission, and they showed the rest of the Marine Corps that on the battlefield aviation could make a difference—sometimes *the* difference between victory and disaster.

All the elements for an air arm that was an integral part of the Marine Corps with a vital role in carrying out the Marines' mission were developed during the 1920s. It remained for Marine aviators in the next decade, under the shadows of depression and an impending Second World War, to bring their service to maturity and point it toward the great struggles and triumphs of the 1940s.

Marine Fokkers on the landing strip at Ocotal, about 1929. (Marine Corps Photo 515413).

CHAPTER IV

MARINE AVIATION COMES OF AGE, 1930–1940

Impact of the Great Depression

For Marine aviation, as for every element of the United States armed forces, depression-induced budget reduction was the dominant fact of the early 1930s. Marine aviation since its beginnings had operated under austere circumstances; its leaders now learned the truth of that old adage, "things could be worse." For 10 years after 1929, and especially in 1930, 1931, and 1932, appropriations for the military sank to survival level, and Marine aviation stood low on the priority list for distributing what funds Congress did allocate.

Marine aviation began a series of cost-cutting reductions, redeployments, and reorganizations. Abandoning the lighter-than-air field, the Marines abolished their balloon squadron (ZKO–1M) at Quantico on 31 December 1929 and distributed its personnel among their other aviation units on the east coast. The following August, they disbanded their lighter-than-air detachment at Great Lakes Naval Training Station. During April 1931, they broke up one observation squadron (VO–10M) at San Diego and transferred its aircraft and personnel to the remaining one (VO–8M) at that station. At Quantico, they merged the aircraft and personnel of two fighting squadrons into one (VF–9M). These changes reduced the administrative cost of operating the aircraft of these units without reducing the total number of aircraft in operation.[1]

In response to both budgetary pressures and to a new mood of isolationism in Congress, Marine aviation liquidated most of its overseas commitments during the early 1930s. On 26 February 1931, the squadron stationed at Sumay, Guam, was withdrawn to the United States. A month later, it was dissolved, its personnel going to other aviation units and its materiel and equipment reverting to the Navy's Bureau of Aeronautics.

Late in 1932, in response to the re-establishment of public order in Nicaragua and to a Congressional ban on the expenditure of any additional military appropriations to support forces in that country, the Marine air units left Nicaragua along with the rest of the Marine brigade. At the end of 1932, Marine aviation had only one remaining overseas commitment—Haiti, where one squadron (now designated VO–9M) continued to provide logistic support for ground forces while conducting routine training. This last commitment came to an end in August 1934 when VO–9M left the island and joined the air group at Quantico.[2]

Aviation and the Fleet Marine Force

While the Depression years brought budget cuts and economy drives to Marine aviation, they also brought a final reorganization and definition of mission. The Marine Corps, with its overseas commitments reduced to a minimum during the early 1930s, undertook a major review of its place in United States strategy. In the course of that review, a debate between two schools of thought within the Corps reached its climax. One faction argued that the Marine Corps should remain a small "Army" capable of performing any mission that the Army could, but on a limited scale. Opposed to adherents of this "jack of all trades" concept were those who believed that the Marine Corps should concentrate on one specialized function—amphibious warfare in co-operation with naval forces with its major objective the seizure of advanced bases for the fleet.

On 8 December 1933, the formation of the Fleet Marine Force (FMF) signalled the triumph of the amphibious warfare advocates. The FMF, drawn from the "force of Marines maintained by the Major General Commandant in a state of readiness for operations with the Fleet," would replace the old East and West Coast Expeditionary Forces. It would be an integral part of the

A Curtiss Hawk flown by Captain Arthur H. Page won the Curtiss Marine Trophy Race at NAS Anacostia on 31 May 1930. This aircraft was modified for racing purposes. (Photo courtesy Major John M. Elliott, USMC, Ret.).

This F6C-4 of Fighting Squadron 10, about 1930, at San Diego has a cowling fitted over the exposed cylinders of its radial engine. (Marine Corps Photo 530812).

Advent of the Boeing F4Bs. A F4B–3 used as the Headquarters Marine Corps command plane in 1933. (Marine Corps Photo 529745).

Last and best of the Boeing biplanes, an F4B–4 of VF–9M in 1935. (Marine Corps Photo 515228).

F4B–3 of Bombing Squadron 4 (VB–4M) in flight in 1935. Equipped for dive-bombing, this aircraft had a bomb rack under the fuselage. (Marine Corps Photo 529974).

F4B–4s of VF–9M line up at Brown Field, Quantico in 1935. (Marine Corps Photo 528314).

fleet, under the operational control of the fleet commander. The Commandant of the Marine Corps retained operational control of units and personnel not attached to the FMF, and he had administrative authority over all Marine personnel and was responsible for the conduct of training. The Commandant also had charge of research and development of doctrine, techniques, and equipment for amphibious warfare.[3]

As initially organized during 1933–1934, the FMF consisted of a regiment of infantry, two batteries of 75mm pack howitzers, one battery of 155mm guns, and one battery of .50 caliber antiaircraft machine guns. The air squadrons of the former East and West Coast Expeditionary Forces were incorporated into the FMF as Aircraft One, located at Quantico, and Aircraft Two, at San Diego. These squadrons, in the words of the Major General Commandant, "form an integral part of the Fleet Marine Force and are organized for the support of that force in its operations with the fleet." Only three squadrons were not attached to the FMF—two which were deployed on board carriers and the one remaining in Haiti. The latter unit joined Aircraft One upon its transfer to Quantico.[4]

Besides organizing the FMF, the Marine Corps began to distill the lessons of long study and years of practical experience into a unified doctrine for the conduct of amphibious operations. During late 1933 and early 1934, the instructors and students at the Marine Corps Schools, in consultation with officers from Headquarters Marine Corps and the FMF, drew up the *Tentative Landing Operations Manual*. This document, published by the Navy Department in 1935, laid out in detail the principal steps for conducting an amphibious assault. The concepts of command relationships, organization, fire support, assault tactics, ship-to-shore movement, and logistics outlined in the manual and refined in edition after edition were tested and improved in fleet exercises during the 1930s. In World War II, they guided Marines to their hard-won Pacific victories.

The aviation section of this famous manual was written by a group of Marine fliers headed by Captain Harold D. Campbell.* It discussed the role of Marine aviation in terms that echoed Cunningham's writings of the early 1920s. It recognized Navy and Marine aircraft, along with naval gunfire, as the sources of fire support for an opposed beach landing, and it declared that an air superiority of at least three to one in the landing areas was a fundamental prerequisite for success.

The *Tentative Manual* listed the functions of aviation at every stage of an amphibious landing—long-range reconnaissance, providing fighter cover over transports and landing craft, knocking out enemy airfields and artillery positions, neutralizing beach strongpoints, artillery spotting, and close support of advancing troops after the beachhead was secured. As had Cunningham, the manual emphasized the importance of communication between aircraft, ships, and ground units and urged that all airplanes be equipped with two-way radios.

While the manual assumed that both Navy and Marine aircraft would be involved in any amphibious assault, it urged that Marine air units take a large part and advocated the assignment of a carrier for their exclusive use. In the *Tentative Landing Operations Manual*, Marine aviation achieved recognition as an integral and vital element in the exeution of the Marine Corps' primary mission, and its functions were defined with sufficient precision to guide organizational and training efforts.[5]

In line with the manual's conclusions, the General Board of the Navy in 1939 summed up the mission for which Marine aviation was to prepare and in fact long had been preparing:

> Marine aviation is to be equipped, organized and trained primarily for the support of the Fleet Marine Force in landing operations in support of troop activities in the field; and secondarily, as replacements for carrier-based naval aircraft.[6]

Colonel Turner did not live to see the air arm he headed for so many years achieve this recognition. In 1931, he made an inspection flight to Haiti in a new Sikorsky amphibian. After a normal landing at Gonaives, Haiti, the aircraft rolled into some soft sand into which the left landing gear sank two feet. Turner jumped from the plane to survey the damage. As he went under the propeller, which was still turning, he forgot to allow for the list of the airplane, and the propeller struck him in the side of the head and killed him. Only 49 years old when he died, Turner had been the first Marine aviator in line for promotion to brigadier general.[7]

Major Roy S. Geiger succeeded Turner as head of the Aviation Section. At this time, the senior Marine airman by rank was Major Ross Rowell, who had led the dive bombing attack at Ocotal, but Geiger had joined aviation five years

---

* The other Marine drafters of the section were First Lieutenants Vernon E. Megee, William O. Brice, Pierson E. Conradt, and Frank D. Wier. (Megee comments).

Colonel Thomas C. Turner was killed on 28 October 1931 at Gonaives, Haiti, when he stepped into the propeller of a Sikorsky RS–1 similar to this one. (U.S. Naval Air Station, Quantico, Photo 299).

before Rowell and had been senior squadron commander with the 1st Aviation Force while Rowell had not received his wings until November 1922. By experience, then, Geiger could claim seniority, and the Major General Commandant put experience ahead of rank in choosing a new chief of aviation. Geiger served until 30 May 1935, participating in some of the conferences at which the *Tentative Manual* was drafted. He then went on to other assignments.

In World War II, Geiger would command the 1st Marine Aircraft Wing during the battles of Guadalcanal and become successively the first Marine aviator to command an amphibious corps and the first Marine to command an army (the Tenth on Okinawa). In 1945, with the rank of lieutenant general, he would command Fleet Marine Force, Pacific.

Geiger's successor as head of aviation, Major Rowell, served until 10 March 1939. During Rowell's tenure, the position of Marine aviation at headquarters underwent a change long sought by its directors. In 1935, the same year that Rowell succeeded Geiger, the Aviation Section was separated from the Division of Operations and Training and placed directly under the Major General Commandant. Then on 1 April 1936, the section achieved full-fledged division status with Rowell, now a colonel, as its first director. As Director of the Division of Aviation,

Colonel Rowell advised the Major General Commandant on all aviation matters and served as liaison officer between Marine headquarters and the Navy's Bureau of Aeronautics (BuAir), upon which Marine aviation still depended for aircraft, equipment, and supplies.[8]

The new status of the Director of Marine Aviation increased the effectiveness with which Colonel Rowell and his successors could plan the development of the Marine air arm and defend its interests in service councils. Through access to the Commandant, the Directors of Aviation could determine what the Marine Corps expected from its aviation component. Through liaison with BuAir, they could ascertain what the Navy required of the Marine air arm and what assets they could obtain to meet the demands. As fleet exercises under the new amphibious doctrines raised problems of aviation command and responsibility, independent Directors of Marine Aviation, dealing directly with the Commandant and the Chief of Naval Operations, could resolve most of the controversies by establishing more precise definitions of responsibility.

Within the framework of Aircraft One and Aircraft Two, FMF, Marine squadrons underwent various redesignations and reorganizations. Always, the direction of these changes was toward more complete commitment to the FMF and to the Marines' missions in support of it. In

Marine aviation joins the carriers. A Vought O2U-2 of VS-14M on the deck of the USS Saratoga in November 1931. The arresting hook can be seen underneath the fusilage. (Marine Corps Photo 529593).

A line of Vought SU-4s of VO-8M. In the 1930, observation planes also began to be called "scout" planes. (Marine Corps Photo 517614).

Vought SU-2 of VS-15M in 1936. Aircraft of this and similar types flew observation missions for the Marines during the Thirties. (Photo from Museums Branch Activities, Quantico).

1934, the two squadrons (VS–14M and VS–15M) which had been stationed on board aircraft carriers since 1931 were disbanded. Reorganized as VO–8M, their aircraft and personnel joined Aircraft Two at San Diego. Meanwhile, VO–9M from Haiti joined Aircraft One at Quantico. These reorganizations left Marine aviation totally committed to the FMF.

In January 1935, Aircraft One consisted of one headquarters squadron (HS–1M), one service squadron (SS–1M), two observation squadrons (VO–7M and VO–9M), one fighting squadron (VF–9M), and one utility squadron (VJ–6M). Aircraft Two at the same time contained a headquarters squadron (HS–2M), a service squadron (SS–2M), an observation squadron (VO–8M), a bombing squadron (VB–4M), and a utility squadron (VJ–7M).[9] Further reinforcing its integration with the fleet, Aircraft Two early in 1935 was placed under the direct authority of the Commander-in-Chief, U.S. Fleet, and further assigned to Aircraft, Battle Force, U.S. Fleet. Under this command arrangement, which prevailed during most of the decade, Aircraft Two spent much time operating from carriers.

In 1936, the neat organizational structure of Aircraft One was disrupted when VO–9M deployed to St. Thomas in the Virgin Islands,

where it operated as an independent unit of the FMF, separate from Aircraft One. The following year, Aircraft Two received a new fighting squadron, VF–4M, and the Marines renumbered all of their squadrons to conform to a new Navy numbering system.* Late in the same year, to simplify accounting and administrative procedures and bring them into line with those of the Navy, the Marines redesignated their non-flying squadrons to differentiate them from the mobile organizations. In Aircraft One and Two, headquarters squadrons were redesignated base air

---

*The new system was as follows:

		Old	New
Aircraft One:	HQ	HS–1M	Same
	Service	SS–1M	Same
	Observation	VO–7M	VMS–1
	Fighting	VF–9M	VMF–1
	Bomber	VB–6M	VMB–1
	Utility	VJ–6M	VMJ–1
Aircraft Two:	HQ	HS–2M	Same
	Service	SS–2M	Same
	Observation	VO–8M	VMS–2
	Fighting	VF–4M	VMF–2
	Bomber	VB–4M	VMB–2
	Utility	VJ–7M	VMJ–2
St. Thomas, V.I.	Observation	VO–9M	VMS–3 [10]

The utility squadron (VJ–7M) of Aircraft Two lined up for inspection at San Diego in 1933. The aircraft in the foreground are N2C–2s, with a Fokker tri-motor at the far end of the line. (Marine Corps Photo 528144).

A Vought SU–2 of VO–9M at Bourne Field, St. Thomas, Virgin Islands. The squadron was stationed here beginning in 1936. (Marine Corps Photo 529595).

An F3F–1 of VF–4M. This was the first Marine fighter with retractable landing gear. (Photo from Museums Branch Activities, Quantico).

The Marines' first all-metal monoplane fighter, the Brewster F2A–3 "Buffalo." (Marine Corps Photo 304388).

Last of the Grumman biplanes, an F3F–2 with closed cockpit and three-blade propeller in 1938. (Marine Corps Photo 525776).

An F2A "Buffalo" taxiing. This aircraft, the most advanced in Marine hands, quickly became obsolete in World War II. (Photo from Museums Branch Activities, Quantico).

detachments while service squadrons became headquarters and service squadrons. Each of these units was attached to a naval air station and controlled by its commanding officer. Additional base air detachments were formed at St. Thomas and Parris Island.

In May 1939, the East and West Coast air groups underwent a final redesignation. At that time, Aircraft One became 1st Marine Aircraft Group (1st MAG) and Aircraft Two became 2d Marine Aircraft Group (2d MAG). While administratively part of the FMF, the 2d MAG continued to be attached to the U.S. Fleet's Aircraft, Battle Force for carrier operations and training.

As early as 1920, Marine aviation organization had provided for a wing* headquarters under which the squadrons would operate. However, until 1938, no wing had been formed. With the attachment of most Marine squadrons to the FMF, interest in the creation of a wing revived. As proposed in October 1938, the wing headquarters would consist of a commander and staff at the brigade level who would be responsible directly to the FMF commander, or the Navy Battle Force commander when under Navy operational control, for the employment and training of the assigned Marine air units. This proposal received the endorsement of the Commander-in-Chief, U.S. Fleet (CinCUS), who envisioned the wing commander as a member of the staff of the Commander, Aircraft, Battle Force, directing Marine squadrons under that officer's control. The FMF commander also favored the proposal as providing a commander and staff with whom his headquarters could work on planning and training. Also, the wing headquarters could take operational control of the two aviation groups, if both ever were concentrated under one FMF commander.

With the plan for a wing headquarters apparently approved by both Navy and Marine authorities, arrangements were made to activate it on 1

July 1939. A conflict developed, however, between the FMF commander and the Aircraft, Battle Force commander over the precise degree of control each would exercise over the wing. After a year of correspondence, the Commandant and CinCUS finally resolved the difference in favor of the FMF commander, placing the wing firmly within the FMF. The headquarters finally was activated in July 1941, but controversy continued over the composition of the wing as a tactical operating force. This issue remained unsettled on 7 December 1941.

Men and Machines, 1930–1940

In spite of the budget cuts of the 1930s, the manpower of Marine aviation slowly increased. In 1935, the Marine Corps had 147 officers on aviation duty, including 110 pilots, and 1,021 enlisted men. By 1939, the numbers had increased to 191 officers, 173 of them pilots, 19 warrant officers, of whom 7 were pilots, and 1,142 enlisted men. The same gradual upward trend continued into 1940.[11]

Marine air personnel in 1939 included besides the regulars, 56 aviation cadets. These cadets came from the Marine Aviation Reserve, which continued to grow and prosper throughout the 1930s.[12] During 1931 and 1932, defying the worst years of the Depression, the Marines commissioned 11 new reserve squadrons—three service, four observation, two fighting, one scout, and one utility. They added two more later in the decade. Often ill-paid or unpaid (the appropriation for reserve aviation fell as low as $700,000 per year), some Marine air reservists paid their own expenses at drills and encampments rather than forego their training.

In 1935, new legislation strengthened the reserve. Public law Number 37, approved on 15 April 1935, created the grade of aviation cadet in the Marine Corps Reserve and provided for the appointment, instruction, and pay of the cadets and for their commissioning as second lieutenants, USMCR, upon satisfactory completion of training.[13] The first list of candidates for cadet appointments included such great World War II names as Gregory ("Pappy") Boyington (who initially failed to qualify) and Robert E. Galer, a 13-plane ace and Medal of Honor winner.

The Naval Reserve Act of 1938, which also applied to the Marine air reserve, provided for increased pay, disability benefits, paid retirement, and other advantages for reservists and, in the words of Marine aviation historian Captain

---

* Definition of the term "wing" in Marine aviation organization has undergone confusing changes since World War I, as have the definitions of and relations between the wing's subordinate groups and squadrons. By 1938, the terminology had evolved close to the modern usage. That is, the Marine aircraft wing was supposed to command an as yet undetermined number of groups which in turn were composed of varying numbers of squadrons. The exact composition of the wing was then and remains today both variable and controversial. Unlike the Air Force wing, which normally consists of groups and squadrons of a single aircraft type, a Marine aircraft wing always has been composed of groups of fixed-wing aircraft of all types and beginning with Korea also included helicopters.

Curtiss SOC–3s of Observation Squadron Two (VMS–2), part of Aircraft Two, Fleet Marine Force in 1933. (Marine Corps Photo 517613).

In 1938, Vought SB2U–1 "Vindicators," all-metal monoplane scout bombers, brought the observation elements of Marine aviation into the same performance range that the F2A did the fighter elements. (Marine Corps Photo 529317).

Edna Loftus Smith, "really set the mood for the Reserve as it exists today." Further legislation in 1939 permitted the promotion to first lieutenant of reserve second lieutenants who had served as such for three years and passed an examination.

At the end of fiscal year 1938, the Marine Aviation Reserve consisted of 15 student aviators training at Pensacola, 10 inspector-instructors and 34 enlisted men on active duty at reserve aviation bases, and 109 officers and 575 enlisted men on inactive duty, plus 63 cadets on active service at Quantico, San Diego, and Pensacola. Many of the reserve units by this time contained manpower of high quality. Major Karl S. Day,* for example, commander of the reserve squadron at Floyd Bennett Field, authored the first standard textbook on instrument flying and radio navigation. Most of Day's pilots, like Day himself, who worked for American Airlines, held jobs in the airline industry and were "keenly interested in what they were doing." Candidates for enlisted billets had to go through a probation period:

> You come out there and work Saturdays and Sundays and do the dirty work, sweeping hangars and stuff like that, and then if you are pretty good at it, maybe six months later you get a chance to enlist as a buck private. That was the kind of outfits these were. If you have material like that to work with, you can do a lot of things.[14]

After 1935, aviation reserve units routinely took their two weeks of active duty every year for training. Frequently during these periods, they conducted joint exercises with ground Marine reserve units, thereby improving their ability to work with regular Marine aviators if necessary in close support of troops.

For both regular and reserve Marine airmen and ground crewmen, the training cycle established in the 1920s continued into the early 1930s with few major changes. Beginning aviators continued to earn their Navy wings at Pensacola, qualifying first in seaplanes and then in landplanes. Then they went on to Navy, Army, and Marine Corps schools and bases for advanced flight and tactical training. Some Marine aviators also took academic work at the Chemical Warfare School, the Naval War College, Harvard University, and the California Institution of Technology. Enlisted men received instruction at various service technical schools.[15]

Among the aviators at Quantico and San Diego, there continued to prevail the individualistic, often undisciplined atmosphere of the 1920s. A Marine squadron commander in the early 30s, a veteran recalled:

> . . . was not a . . . commander in the sense of Courts and Boards; he had a first sergeant who took care of the service record books, and then a collection of pilots who ran around doing what they pleased. At a place like Quantico there was only one commanding officer, and . . . he had all power of—let's say final power over personnel matters, he had all authority there was over materiel matters, he controlled the station. And the squadron commanders were just people who flew airplanes, flew the number one airplanes, everybody else followed along. The squadron commanders exercised no command at all.[6]

Under this system, "The pilots, the squadrons, were loosely controlled mobs . . . but they were all good airmen, they could all fly like mad." At annual gunnery and bombing exercises, "The umpiring and observing was lax, loose. . . ."[17]

This atmosphere began to change with the start in 1931 of carrier training for Marine squadrons. In that year, VS–14M under Captain William J. Wallace began operations from USS *Saratoga* and VS–15M under First Lieutenant William O. Brice joined USS *Lexington*. Actually detachments rather than squadrons, each of these units consisted of eight aviators and 36 enlisted men and operated six planes. During the three years that these units flew from the carriers, which were based on the Pacific Coast, two-thirds of the Marines' total complement of aviators served with one or the other of them for training. In their shipboard tours, these Marine pilots practiced carrier takeoffs and landings, and they underwent intensive training in gunnery, formation flying, aerial tactics, and communications, training checked periodically by thorough tests and inspections.

This curriculum, standard for Navy fliers at the time, would appear loosely organized to a modern naval aviator, but it seemed highly formalized to the Marine pilots. In the words of one, Edward C. Dyer, it was:

> . . . a rude awakening. . . . There was no monkey business whatsoever. In the first place we were handed a doctrine, a book, a guide, that told us how the squadron should be organized. . . . We had a commanding officer, an executive officer, a flight officer, an engineering officer, a materiel officer, and so on, and the duties of each officer were all spelled out. . . . The organization and operation of the

---

* Recalled to active duty in 1940, Day went on to a distinguished Marine aviation career in World War II and after the war remained active in reserve affairs. Before retiring with the rank of lieutenant general in the Marine Corps Reserve, he played a major part in legislative battles for the survival and growth of the reserve and served from 1953–1956 as President of the Marine Corps Reserve Officers' Association. He died on 19 January 1973.

SB2U–3 in flight. This aircraft was classed as a scout bomber and could take off from carriers or be launched from a ship's catapult. (Marine Corps Photo 306304).

Marines received new transports during the 1930s, including this Curtiss-Wright R4C–1 "Condor" transport in 1937 which had a crew of two and could carry 10 passengers. (Marine Corps Photo 517615).

squadron was definitely controlled. The aircraft were issued by the Air Battle Force material people. They would . . . give us the airplanes; we would then have to maintain them. But these fellows would arrive and inspect. They'd swoop down from the staff and take a look at your airplanes just to see if you were maintaining them in a satisfactory condition. . . . All of our material was requisitioned and accounted for. We were required to follow a training syllabus. We had so many hours of gunnery, so many hours of navigation, so many hours of radio practice, so many hours of formation flying, so many hours of night flying, and we jolly well had to do it. . . .[18]

Aviators returning from tours with the carriers introduced new standards of professional performance to the squadrons at Quantico and San Diego, and commanders like Colonel Rowell worked to improve training and tighten discipline. In 1938–1939, the FMF instituted a four-phase training plan intended to achieve "coordinated and progressive training of all units, in order to prepare the command for immediate operations with the United States Fleet."

Marine aviation had an assignment in each phase, beginning with individual gunnery practice and then progressing to squadron tactics and formation flying, navigation, night flying and instrument flying, and practice in supporting ground troops. In the final phases, all squadrons of the 1st and 2d MAGs joined the ground elements of the FMF in large-scale fleet landing exercises. As a result of these influences, Marine aviation by 1940 was becoming a fighting organization oriented toward its principal mission rather than a random collection of pilots and aircraft.[19]

Marine aviators in the 1930s trained and operated with aircraft of steadily improving performance, mission capability, and reliability. Around 1932, they began receiving fighters of the Boeing F4B series, the famous Boeing "Bipes." With all-metal fuselages in the later models and wood framed, fabric-skinned wings, these sturdy biplanes served both as fighters and dive bombers. The latest and best of the series, the F4B–4, was armed with one .30 and one .50 caliber machine gun and could carry two 116-pound bombs in wing racks. With its 550-horsepower Pratt and Whitney radial engine, it could reach a top speed of 184 miles per hour and a service ceiling of 26,900 feet. It had a cruising range of 350 miles which could be extended to 700 by fitting an external fuel tank under the belly. Pilots found the F4B–4 easy to fly; it maneuvered readily and responded quickly to the controls. A Marine aviator remembered the F4B–4 as "the *one* airplane which made the pilot feel that he himself was flying—not just riding in a machine."

In the late 30s the Grumman F1F, F2F, and F3F series, perhaps the ultimate in biplane fighter design and performance, supplanted the F4Bs. All-metal in construction except for fabric-covered wings and control surfaces, these small (28-foot wingspread) airplanes boasted such features as enclosed cockpits and retractable landing gear.

The final plane of the series received in quantity by the Marines, the F3F–2, had an 850-horsepower Wright Cyclone radial engine and could reach a top speed of 260 miles per hour. Its service ceiling was about 32,000 feet, and it had a range of 975 miles at a cruising speed of 125 miles per hour. Pilots unfamiliar with its retractable landing gear, which had to be raised and lowered by a hand-cranked gear and chain, made numerous wheels-up landings in the F3F–2, but the sturdy machines usually escaped from these mishaps with little damage other than bent propellers, torn skins, and dented cowlings.

Finally, in 1939, the Marines received their first all-metal monoplane fighter, the Brewster Aeronautical Corporation's F2A "Buffalo." This craft, faster and more heavily armed than its predecessors, itself would become obsolete before it entered combat as the pressures of World War II accelerated airplane development.

Evolution of other aircraft types paralleled that of fighters. For observation planes, the Marines throughout most of the decade used the Vought SU–1 through 4 series and the Curtiss SOC–3. All of these were single-engine, two-seater biplanes with top speeds of around 160 miles per hour. In the late 30s, these gave way to the Vought SB2U–1 and SB2U–3 "Vindicator," a single-engine, two-seater, all-metal monoplane. For dive bombing, Marine aviators in the mid-30s began using the Great Lakes BG–1, a big, rugged biplane which would remain in service until replaced (in 1941) by the monoplane Douglas SBD series.

Transport aircraft also steadily improved. In the early part of the decade, the Marines continued to use the Ford and Fokker tri-motors that had proven their worth in Nicaragua. In 1935, they received two new models of the Ford tri-motor, RR–4s, each powered by three 450-horsepower Wasp radial engines. In June 1934 and November 1935, they supplemented their Fords and Fokkers with two Curtiss R4C–1 "Condors," twin-engined biplanes. The DC–2, designated the Douglas R2D–1, a low wing, twin engine, all-metal transport and the ancestor of

In 1935 in the Douglas R2D–1 the Marines made the acquaintance of the ancestor of the World War II "Gooney Bird" and crossed the threshold of modern air transport capability. (Photo Courtesy of Marine Corps Museum, Quantico).

The 1934 flight line of VO–8M at NAS San Diego. The aircraft are Vought O3U–6 observation planes and Curtiss R4C–1 "Condors." (Marine Corps Photo 530257).

the World War II "Gooney Bird," entered Marine aviation in 1935. With these few large transports and several smaller twin-engine utility machines, Marine aviators in the 1930s gained airlift experience which would prove invaluable during the early days of World War II in the Pacific.[20]

During 1932, their last year in Nicaragua, Marine aviators at Managua tested their first vertical takeoff and landing, rotary-winged aircraft—a Pitcairn autogyro,* one of three experimental models which the Navy had purchased from the manufacturer. On test flights around Managua, the ungainly craft, with its overhead rotor and stubby wings, attracted great attention from the Nicaraguans who developed a proprietary fondness for it. The Marines liked it less well. While the machine could take off and land in a space smaller than that required by conventional aircraft of the day, it was difficult to fly and could carry a payload of no more than 50 pounds. In a report to Headquarters Marine Corps dated 22 November 1932, the aviators who had tested the autogyro concluded that it had no expeditionary use beyond limited reconnaissance and passenger-carrying functions. For the time being, and in fact until after World War II, Marine Corps aviation would continue to rely on fixed-wing aircraft.[21]

Operations, 1930–1940

Marine air operations during the 1930s reflected the increasing capabilities and enhanced sense of mission and purpose of the aviation service. While the air races, exhibition flights, and formation flyovers of the 20s continued into the new decade, they took an inferior place on the list of priorities to fleet problems, landing exercises, and the annual qualification for record in aerial gunnery and bombing.

Air races continued to be popular during the 30s, and Marines continued to compete in them.

On 31 May 1930, Captain Arthur H. Page won the Curtiss Marine Trophy Race held at Anacostia Naval Air Station. Flying an F6C–3 landplane modified and equipped with pontoons for the event, he completed the five laps around the 20-mile course at an average speed of 164 miles per hour.

An enthusiastic competitor, Captain Page did not content himself with success in the Curtiss Trophy race. He also established a distance record for "blind"** flying by making a 1,000-mile instrument flight in an O2U–1 Corsair from Omaha, Nebraska, to Washington, D.C. In September 1930, Captain Page was on his way to his third success of the year, leading all entries through 17 of the 20 laps of the Thompson Trophy Race in Chicago, when he was overcome by carbon monoxide leaking into his cockpit, crashed, and died in the wreck.

Captain Page's death did not end Marine fliers' efforts to publicize their service and educate the American people about the various functions of military aviation. Marine pilots participated in dive bombing exhibitions, parachute jumps, and formation flights. They competed in the National Air Races at Cleveland, Ohio, and the American Air Races at Miami. Carrying the banners of Marine aviation to Canada, they took part in the Toronto Flying Club Pageant. In a continuing effort to prove by performance the value of their service, Marine aviators for a while carried the air mail between Washington, D.C., and Camp Rapidan, Virginia, a frequently-used conference site for government leaders.[22]

These activities, however, declined in importance during the 1930s compared to the serious work of training Marine aviation to support the FMF. With no overseas commitments to divert their energies, the Marine squadrons' major operations of the decade all were related to preparation for amphibious warfare. The 1st MAG at Quantico, treated as part of the 1st Marine Brigade at that station, centered its operations on preparation to support either the FMF or fleet aviation. On the Pacific Coast, the Marines disbanded VS–14M and VS–15M in 1934 and transferred their personnel and equipment to Aircraft Two (2d MAG) at San Diego. Thereafter, all squadrons of 2d MAG usually were attached to Aircraft, Battle Force, and spent much of their time flying from carriers while participating in exercises

---

* The autogyro, like the helicopter, derived its lift from an overhead rotor, but unlike the helicopter, it did not apply engine power to the rotor in flight. A single engine powered both the rotor and a front propeller. In taking off, the pilot first used a clutch to connect the engine to the overhead rotor. After bringing it up to takeoff speed, he switched power to the front propeller, leaving the rotor turning freely. The machine then was supposed to lift into the air after a short takeoff run and fly with the spinning rotor and stubby wings providing air lift. The autogyro could take off and land almost vertically, but it could not hover as can a helicopter.

** The term was used at that time to denote control of an aircraft by a pilot using aircraft instruments only without visual reference outside the aircraft.

with the Pacific Fleet. On both coasts, the primary objective of all FMF aviation remained, in the Commandant's words: "the close support of troops in a landing and during the operations subsequent thereto." [23]

In 1935 and every year thereafter until the United States entered World War II, the 1st and 2d MAGs took part in the fleet landing exercises in which the methods of amphibious warfare were tested and refined. [24] Early in 1935, 12 aircraft from 1st MAG joined the 1st Marine Brigade in Fleet Exercise Number One at Culebra, Puerto Rico. This squadron experimented with techniques for spotting the fall of naval gunfire in shore bombardments and practiced bombing and strafing beach targets representing defense installations. To their surprise, Marine pilots found low-altitude bombing more effective than dive bombing in these attacks.

From 4 January to 24 February 1936, the entire 1st MAG, over 50 planes, again supporting the 1st Marine Brigade, participated in Fleet Landing Exercise Number Two at Culebra. For a month, while the infantry made eight separate beach landings, the aircraft laid smoke screens, bombed and strafed beach targets, spotted for naval gunfire, and flew reconnaissance and photographic missions. Marine aviators learned this time that smoke screens laid from the air disrupted rather than protected formations of assault boats by reducing the boat crews' range of vision. This series of maneuvers, like others that followed, also convinced Marine aviators that they needed a specialized attack plane properly to conduct their mission of close support. The current practice of using fighter and observation machines for this purpose, in one Marine pilot's words, "interferes materially with the normal missions of these types, and is at best a makeshift expedient."

In 1937, for the first time, both Marine air groups, mustering between them 83 aircraft, operated together as a combined force. For this event, 1st MAG flew across the continent to join 2d MAG in Fleet Landing Exercise Number Four, held between 27 January and 10 March 1937 at San Clemente Island, California. This major exercise involved both the 1st and 2d Marine Brigades, as well as a provisional Army amphibious brigade. In this series of maneuvers, Navy carrier planes did all the gunfire spotting, and the Marines concentrated on general reconnaissance, observation, and attacks in support of ground troops. Once again, Marine aviators came away from the maneuvers convinced they

needed a specialized attack aircraft. Similar fleet exercises followed every year through 1941.

In all of these exercises in the 1930s, Marine aviators were supposed to be improving their ability to give close support to infantry in the ground battle. The decade ended, however, with major operational problems still unsolved and with the concept of close air support itself as yet ill-defined.

Marine fliers found their close support efforts hindered rather than helped by their new fast, higher-flying aircraft. Pilots in the open cockpits of slow-moving DH–4Bs and comparable machines usually could locate friendly and enemy positions relatively easily by sight and sound, but aviators of the 30s, often riding in closed cockpits, swept across the lines too quickly to orient themselves. Also, the Marine fliers of the 30s, who had specialized in aviation from the beginning of their military careers, lacked the familiarity with ground tactics possessed by aviators of the older generation, many of whom had transferred from the infantry.

By the end of the decade, both ground and air Marines realized that the solutions to these problems lay in improved radio communication, simplified and mutually understood systems for locating ground targets, and still more intensive joint training, but the implementation of these measures remained incomplete at the outbreak of the war with Japan. [25]

Partly as a result of these continuing practical difficulties, the Marine Corps Schools as late as 1940 defined the role of aviation in supporting infantry in cautious terms:

> When aviation is acting in close support of the ground forces, its striking power should be used against [only] those targets which cannot be reached by the ground arms, or on targets for which ground weapons are not suitable or available. In almost all ground situations there are vital targets beyond the range of weapons of ground arms which can be powerfully dealt with by attack aviation. Therefore, the use of attack aviation to supplement the firepower of ground arms is generally discouraged as it may result in the neglect of more distant, and perhaps more vital objectives. As a general rule, attack aviation should be used in lieu of artillery only when the time limit precludes the assembly of sufficient artillery units to provide the necessary preparation, and when such absence of artillery may involve failure of the campaign as a whole. [26]

Nevertheless, while the question of close air support remained the subject of debate, by the end of the 1930s the Marine air-ground team had moved a long distance from concept toward reality. The conduct of amphibious warfare, including aviation's part in it, had been formu-

F3F–2s of VMF–2 flying in formation in 1938. (Marine Corps Photo 515234).

A Pitcairn autogyro of the type Marines tested in Nicaragua in 1933. While a forerunner of rotary-wing and vertical takeoff and landing craft, the autogyro proved ineffective for Marine purposes. (Marine Corps Photo 514902).

SU–2s of VO–8M fly in formation over San Diego in 1933. (Marine Corps Photo 530122).

A Great Lakes BG–1 of Marine Bombing Squadron Two (VMB–2) in 1935. These large, sturdy biplanes were used as dive bombers by Marine aviators in the 1930s.(Marine Corps Photo 529314).

lated into a doctrine which had been tested and validated insofar as could be done within the limitations of peacetime exercises. Marine aviation's command and staff organization had evolved from independent squadrons into a wing completely integrated into the FMF, and large-scale training of air units with ground forces had become routine. Marine air's relationship with the Navy had been clearly defined. In fleet problems and landing exercises, Marine aviation had demonstrated potential ability to perform both its primary mission of supporting the landing force and its secondary mission of reinforcing Navy units on carriers.

Conclusion: Marine Corps Aviation, 1912–1940

Marine aviation began in the years 1912–1917 with a few men experimenting with rickety machines, their concept of an operation being to take off, fly a few miles, and land again with the aircraft still in one piece. As aircraft gradually improved in performance and reliability, and as the Marine Corps, like the other services, slowly committed more men and resources to aviation, a rudimentary organization began to take form, and Marine aviators began to see the outlines of a mission: support of Marine expeditionary forces in seizing and holding advance bases.

In World War I, the first war in which Airpower played a significant part, Marine aviation, like the Corps as a whole, was diverted from its amphibious expeditionary mission and sent into large-scale land combat in France. Denied the opportunity which they eagerly sought to support the Marine brigade, Marine aviators managed to place organized squadrons at the front, and they fought where they were needed. They proved their ability to hold their own in combat against German veterans.

During the 1920s, Marine aviation, although hampered by limited budgets and often outmoded equipment and diverted by the showmanship and headline-hunting of the decade, still moved toward definition of its role in supporting Marine operations. In the air over Haiti, the Dominican Republic, China, and Nicaragua, Marine aviators actively assisted the ground forces, not only in combat but also in reconnaissance, transportation, and supply. By trial and error they worked out basic tactics for close air support. In Nicaragua, by the end of the decade, the Marine air-ground team had become a reality.

Then in the 1930s, as Marine Corps doctrine crystallized and it began to train for its amphibious warfare mission, Marine aviation achieved full acceptance as part of the Fleet Marine Force, as well as developing a secondary capacity to join naval aviation in carrier operations.

In all of these decades, Marine aviators developed versatility. They flew all kinds of missions with all kinds of airplanes. They learned early to make do and do well with what they had. They established a tradition of excellence and adaptability which would be seen again and again, from Henderson Field on Guadalcanal to the frozen hills around the Chosin Reservoir to the monsoon skies of Vietnam.

NOTES

Unless otherwise noted, the material in this monograph is drawn from the following sources: LCdr Reginald Wright Arthur, USN (Ret.), *Contact!* (Washington, D.C.: Naval Register, 1967); Capt Charles W. Boggs, USMC, "Marine Aviation: Origin and Growth," *Marine Corps Gazette*, v. XXXIV, No. 11 (November 1950), pp. 68–75; Maj Edwin N. McClellan, USMC, "Marine Corps Aviation," *Marine Corps Gazette*, v. XVI, No. 1 (May 1931), pp. 11–13, 43–44 and No. 2 (August 1931), pp. 56–59; Martin Caidin, *Golden Wings* (New York: Random House, 1960); RAdm George van Deurs, USN (Ret.), *Wings for the Fleet* (Annapolis: U.S. Naval Institute, 1966); Robert Sherrod, *History of Marine Corps Aviation in World War II* (Washington, D.C.: Combat Forces Press, 1952), hereafter Sherrod, *Marine Air History*; Adrien O. Van Wyen and Lee M. Pearson, *United States Naval Aviation, 1910–1960* (Washington, D.C.: Government Printing Office, 1961), hereafter Van Wyen and Pearson, *Naval Aviation*; Capt Archibald D. Turnbull, USN and LtCdr Clifford L. Lord, USNR, *History of United States Naval Aviation* (New Haven, Conn.: Yale University Press, 1949). All references to the annual reports of the Commandant of the Marine Corps are taken from extracts in the file "Major General Commandant, Annual Reports on Aviation, 1912–1941." All cited material is available in or through the Historical Reference Section, History and Museums Division, Headquarters, U.S. Marine Corps (HRS, HMD, HQMC).

CHAPTER I.
The Beginnings, 1912–1917

Unless otherwise indicated, all material in this chapter on the development of naval aviation is drawn from Turnbull and Lord, *History of US Naval Aviation, passim*, and Van Deurs, *Wings for the Fleet, passim*. This chapter also draws heavily upon material supplied by two commentators on the original draft of the manuscript: MSgt Walter F. Gemeinhardt, USMC (Ret), ltr to Director, History & Museums Div, dtd 14 May 1975, Comment File, Brief History of Marine Corps Aviation (hereafter cited as Gemeinhardt Comments); and Lee M. Pearson, Historian, NavAirSysCom, ltr to Director, History & Museums Div, dtd 26 March 1975, Comment File, Brief History of Marine Corps Aviation (hereafter cited as Pearson Comments).

1 LtCol Kenneth J. Clifford, USMCR, *Progress and Purpose: A Developmental History of the United States Marine Corps, 1900–1970* (Washington, D.C.: HMD, HQMC, 1973), pp. 1–16, hereafter Clifford, *Progress and Purpose*; MajGenCmdt, Annual Report, 1912.
2 Alfred A. Cunningham Biographical File (HRS, HMD, HQMC).
3 Lt A. A. Cunningham ltr to Capt. W. I. Chambers, dtd 25 July 1913, quoted in Caidin, *Golden Wings*, p. 8.
4 McClellan, "Marine Corps Aviation," *Marine Corps*

Gazette (hereafter cited as *MCG*), v. XVI, No. 1 (May 1931), p. 12.
5 Order from MajGenCmdt to 1stLt A. A. Cunningham, dtd 10 May 1912, Cunningham Bio File.
6 Van Deurs, *Wings for the Fleet*, p. 76.
7 Maj A. A. Cunningham, ltr to Mr. Miller, dtd 22 Jan 1931, Cunningham Papers; Maj A. A. Cunningham, ltr to Maj E. W. McClellan, dtd 11 Mar 31, McClellan Papers (Collections Unit, Marine Corps Museums).
8 Bernard L. Smith Biographical File (HRS, HMD, HQMC); Van Deurs, *Wings for the Fleet*, p. 76
9 McClellan, "Marine Aviation," *MCG*, v. XVI, No. 1 (May 1913), pp. 13, 43.
10 Quoted in Caidin, *Golden Wings*, p. 9.
11 The details of these flights can be found in the typewritten Log Book in the Cunningham Papers.
12 Cunningham Bio File.
13 MajGenCmdt, Annual Report, 1913.
14 The Chambers Board's findings are summarized in Turnbull & Lord, *History of US Naval Aviation*, pp. 33–34.
15 Capt Edna Loftus Smith, MCWR, "Aviation Organization in the U.S. Marine Corps, 1912–1945" (Ms in HRS, HMD, HQMC), p. 2, hereafter cited as Smith, "Aviation Organization"; Clifford, *Progress & Purpose*, pp. 18–21; 1stLt B. L. Smith, ltr to CO, Flying School, dtd 10 Mar 1914, Subj: Report of Work Done by the Marine Section of the Naval Flying School... while Temporarily attached to the First Advance Base Brigade, USMC, Culebra, P.R., Brief History of Marine Aviation Comment File (We are indebted to Mr. Lee M. Pearson, NavAirSysCom for furnishing us with a copy of this document from Navy files).
16 Turnbull and Lord, *History of US Naval Aviation*, pp. 41–42.
17 Smith Bio File.
18 Maj A. A. Cunningham, ltr to Mr. Miller, dtd 22 Jan 1931, Cunningham Papers.
19 Cunningham Bio File.
20 Van Deurs, *Wings for the Fleet*, pp. 154–155, discusses Evans' feat and the arguments about looping and spinning in some detail.
21 MajGenCmdt, Annual Report, 1916.
22 *Ibid.*
23 MajGenCmdt, order to 1stLt A. A. Cunningham, dtd 26 Feb 1917, copy in Cunningham Bio File; Smith *Aviation Organization*, p. 2.

CHAPTER II
Marine Aviation in World War 1

Unless otherwise noted, material in this chapter is drawn from the sources listed at the beginning of the notes and in addition from MSgt Roger M. Emmons, USMC (Ret), *The First Marine Aviation Force, 1917–1918: Development and Deployment* (Ms File Copy, HRS, HMD, HQMC); this paper also can be found in *Cross and*

Cockade: The Journal of the Society of World War I Aero Historians, v. VI, No. 2 (Summer 1965), pp. 173–186, and No. 3 (Autumn 1965), pp. 272–292. All page references to this source in this monograph are to the Ms file copy. Material for this section is drawn also from Smith, "Aviation Organization."

1 MGen Ford O. Rogers, USMC (Ret), Interview by Oral History Unit, HMD, dtd 3 Dec 1970 (Oral History Coll, HMD), tnscrpt, p. 25, hereafter Rogers Interview.

2 MajGenComdt George Barnett, ltr to CNO, Subj: Organization of Land Aero Squadron, dtd 27 July 1917, MajGen Comt Barnett, ltr to BGen Squires, Ch Sig Off, US Army, Subj: US Marine Aviation Section for Service in France, dtd 17 Sept 1917; Col J. T. Nance, USA, ltr to MGC, Subj: Instruction of Commissioned and Enlisted Mar Corps Personnel in Ballooning, dtd 28 Sept 1917, all in Roger Emmons Papers (Collections Unit, Marine Corps Museums). The Emmons papers are a compilation of Marine aviation records made by MSgt Emmons, Historian, Marine Corps Aviation Association, who graciously turned over copies of them to the History and Museums Division.

3 LtGen Karl S. Day, USMCR (Ret), Interview by Oral History Unit, HMD, dtd 5 Aug 1968 (Oral History Coll, HMD) tnscrpt, pp. 5–6, 8, hereafter Day Interview.

4 MGen Lawson H. M. Sanderson, USMC (ret), Interview by Oral History Unit, HMD, dtd 14 July 1969 (Oral History Coll), HMD, tnscrpt, p. 3, hereafter Sanderson Interview.

5 Quoted in Div of Reserve, HQMC, *The Marine Corps Reserve: A Short History* (Washington, D.C.: US Govt. Printing Office, 1966), p. 12.

6 Fred T. Jane, *Jane's All the World's Aircraft 1919*, C. G. Grey, ed. (New York: Arco Publishing Co., 1969), p. 452a, hereafter Jane, *World Aircraft 1919*.

7 Gen Christian F. Schilt, USMC (Ret), Interview by Oral History Unit, HMD, dtd 17 and 21 Nov 1969 (Oral History Coll, HMD) tnscrpt, p. 15, hereafter Schilt Interview.

8 Capt A. A. Cunningham, ltr to MajGen Cmdt, Subj: Intvw with Army Signal Corps Officers, dtd 10 Oct 1917, Emmons Papers.

9 Day Intvw, pp. 9–10.

10 CO, 1st Regt, Fixed Defense Force, ltr to MajGen Cmdt, Subj: Organization of Advance Base Unit, dtd 21 Dec 1917, Emmons Papers.

11 The details of this trip, with interesting observations of France at war, can be found in a pocket diary kept by Cunningham and now part of the Cunningham Papers. The diary has been published by the History and Museums Division as *Marine Flyer in France: The Diary of Captain Alfred A. Cunningham, November 1917–January 1918* (Washington, D.C.: HQMC, 1974).

12 Maj A. A. Cunningham, "Value of Aviation to the Marine Corps," *MCG*, v. V, No. 3 (Sept 1920), p. 224.

13 MajGen Cmdt, order to Capt A. A. Cunningham, dtd 11 March 1918, Cunningham Bio File.

14 The evolution of the Northern Bombing Group can be traced in "Summary of Events in the Development of the Northern Bombing Group," a typed compilation of letters, telegrams, and official reports in the Cunningham papers.

15 MajGen Cmdt to CO, 1st Aviation Squadron, Subj: Change of Station, dtd 31 Dec 1917, in Emmons Papers; for the other side of the story, see Day Intvw, pp. 9–11, and MGen Fred S. Robillard, USMC (Ret), Colonel Robillard's Story on Marine Aviation, in MGen Fred S. Robillard File, Personal Papers (Collections Unit, Marine Corps Museums).

16 MajGenCmdt, Orders to Capt R. S. Geiger, Subj: Change of Station, dtd 4 Feb 1918 and Capt. R. S. Geiger, ltr to MajGen Cmdt, Subj: Transfer of Aeronautic Detachment, dtd 9 Feb 1918, both in Emmons Papers.

17 Rogers Intvw, pp. 26–27, 29.

18 Emmons, *First Aviation Force*, p. 14.

19 MSgt Roger M. Emmons, USMC (Ret), *Navy Fliers Transferred to Marine Aviation, Marine Day Wing in France, 1918* (Baltimore, Md.: 1st Marine Aviation Force Veterans' Association, 1971), pp. 1–10; Rogers Intvw, pp. 36–37.

20 Rogers Interview, p. 33. See also Day Interview, pp. 12–14.

21 Captain Mims's letters to Cunningham and Cunningham's letters to Mims are in the Cunningham Papers; these documents cover the entire period when Cunningham was overseas.

22 Maj Edna Loftus Smith, USMCWR, *Marine Corps Reserve Aviation, 1916–1957* (Washington, D.C., 1959), pp. 2, 5. Hereafter Smith, *Reserve Aviation*.

23 MajGen Cmdt, Annual Report, 1918.

24 Maj Alfred A. Cunningham, ltr to Gen Charles G. Long, dtd 31 October 1918, and ltr to Mr. Miller, dtd 22 January 1931, both in Cunningham Papers.

25 Day Interview, p. 25.

26 Emmons, *First Aviation Force*, pp. 35–36; Maj Alfred A. Cunningham, ltr to Mr. Miller, dtd 22 January 1931, Cunningham Papers.

27 Day Interview, pp. 17, 26; Rogers Interview, pp. 35–36, 38, 41, 44. Both of these Marine aviators flew with the British squadrons.

28 Maj Alfred A. Cunningham, ltr to Gen Charles C. Long, dtd 9 November 1918, Cunningham Papers.

29 Maj Alfred A. Cunningham, ltr to Gen Charles C. Long, dtd 12 October 1918, Cunningham Papers.

30 For history and technical details of the DH–4, DH–9, and DH–9A, see W. M. Lamberton, compiler, and E. F. Cheesman, editor, *Reconnaissance and Bomber Aircraft of the 1914–1918 War* (Los Angeles, Calif.: Aero Publishers, Inc., 1962), pp. 13–14, 36–41, 214–215; also Emmons, *First Aviation Force*, p. 35.

31 For details of the raid on Thielt, see Emmons, *First Aviation Force*, pp. 48–53. The Ralph Talbot Papers provide biographical data on Talbot as well as a moving re-creation of the naive gallantry of these World War I aviators.

32 Emmons, *First Aviation Force*, p. 65.

33 Maj Alfred A. Cunningham, ltr to Gen Charles G. Long, dtd 9 November 1918, Cunningham Papers.

CHAPTER III

Advance Toward Maturity, 1919–1929

1 U.S. Navy Dept., *Annual Reports of the Navy Department for the Fiscal Year 1919* (Washington: Government Printing Office, 1920), v. I, p. 41.

2 Smith, *Aviation Organization*, p. 5.

3 Maj Alfred A. Cunningham, "Value of Aviation to the Marine Corps," MCG, v. V, No. 3 (September 1920), p. 222, hereafter, Cunningham, "Value of Aviation."

4 Smith, "Aviation Organization," pp. 7–8, quotes Cunningham's testimony before the General Board. For Cunningham's views on the role of Marine aviation, see Cunningham, "Value of Aviation," pp. 221–233.

5 MajGen Cmdt, Annual Report, 1919, 1920.

6 Unless otherwise noted, all material in this section is drawn from Smith, "Aviation Organization," pp. 5–10.

7 Quoted in Smith, "Aviation Organization," pp. 5–6.

8 For the rivalry between Cunningham and Turner, see Capt Harvey B. Mims, ltr to Maj Alfred A. Cunningham, dtd 7 October 1918, and Maj Alfred A. Cunningham, ltr to Capt Harvey B. Mims, dtd 30 October 1918 and ltr to Mr. Miller, dtd 22 January 1931, all in Cunningham Papers. Also Rogers Interview, pp. 50–51.

9 Maj Alfred A. Cunningham, ltr to MajGenCmdt John A. Lejeune, dtd 23 February 1928, in Cunningham Biographical File.

10 MGen Louis E. Woods, USMC (Ret), Interview by Oral History Unit, dtd 3, 10, 17, 24 May, 27 June, 7 July 1968 (Oral History Coll, HMD), transcript, p. 63, hereafter Woods Interview; Rogers Interview, p. 15.

11 Woods Interview, p. 62.

12 Maj Edwin H. Brainard, USMC, "Marine Aviation—A Lecture," MCG, v. XI, No. 3 (September 1926), p. 192.

13 Ibid., pp. 192–197.

14 Officer in Charge of Aviation, Memo. to: Director of Operations and Training, Subj: Aviation Data for Quarterly Preparedness Report, dtd 18 October 1921, in File MajGen Cmdt, Annual Reports on Aviation, 1912–1940; Smith, Reserve Aviation, pp. 18–19.

15 Woods Interview, p. 63.

16 LtGen Francis P. Mulcahy, USMC (Ret), Interview by Oral History Unit, HMD, HQMC, dtd 11, 14, 15 February 1967 (Oral History Coll, HMD), transcript pp. 79–80, hereafter Mulcahy Interview.

17 Smith, Reserve Aviation, pp. 29–38.

18 For descriptions of Marine air training, see MajGenCmdt, Annual Reports, 1921 through 1929.

19 Woods Interview, p. 28; see also pp. 29–30.

20 Rogers Interview, p. 50.

21 MajGenCmdt, Annual Report, 1927.

22 MajGenCmdt, Annual Report, 1920; William T. Larkins, U.S. Marine Corps Aircraft, 1914–1959 (Concord, Calif.: Aviation History Publications, 1959), pp. 16, 18; Paul R. Matt, compiler, and Bruce Robertson, ed., United States Navy and Marine Corps Fighters, 1918–1962 (Fallbrook, Calif.: Aero Publishers, 1962), pp. 22–24; Rogers Interview, pp. 56–57, describes the fate of the D–7s.

23 Schilt Interview, pp. 35–36; Larkins, USMC Aircraft, pp. 12–13.

24 Officer in Charge of Aviation, Aviation Data for Preparedness Report for Quarter Ending 30 June 1925, dtd 2 July 1925, in "MajGenCmdt, Annual Reports on Aviation, 1912–1941."

25 For a summary of these developments, see Matt and Robertson, Navy and Marine Fighters, p. 31.

26 BGen Daniel W. Torrey, USMC (Ret), Interview by Oral History Unit, dtd 13 August 1968, (Oral Histor Coll, HMD) transcript, pp. 15–16, hereafter Torrey Interview; Matt and Robertson, Navy and Marine Fighters, pp. 32–36.

27 The comment on the Corsair is in Schilt Interview, pp. 57, 74; Matt and Robertson, Navy and Marine Fighters, pp. 37–46.

28 MajGenCmdt, Annual Reports, 1922, 1923, 1924; Smith, Reserve Aviation, pp. 19–20.

29 Rogers Interview, pp. 53–54, 58–63.

30 Sherrod, Marine Air History, p. 22.

31 Woods Interview, pp. 31–36; 1st Sergeant Harry L.

Blackwell, ltr to his mother, dtd 2 May 1923, in Harry L. Blackwell Papers, Collections Unit, Marine Corps Museums.

32 Capt Ralph J. Mitchell, USMC, "The Longest American Airplane Flight," MCG, v. IX, No. 1 (March 1924), pp. 57–64.

33 U.S. Navy Dept., Annual Reports of the Navy Department for the Fiscal Year 1924 (Washington, D.C.: Government Printing Office, 1925), p. 617; "Professional Notes," MCG, v. XIII, No. 2 (June 1928), p. 156.

34 Rogers Interview, p. 58.

35 Sherrod, Marine Air History, pp. 22–23.

36 Col L. H. M. Sanderson, ltr to Capt Warren J. Goodman, USMCR, dtd 1 September 1944, and ltr to the Chief of Naval Operations, dtd 19 August 1946, both in "Aircraft—Bombers" Subject File, HRS, HMD.

37 Capt Garrett H. Graham, USMCR, ltr to BGen Louis E. Woods, USMC, Subj: Origin of Dive Bombing, dtd 22 February 1944; MGen Ross E. Rowell, USMC, Interview on the Origin and Early Use of Dive Bombing Tactics, held in the Aviation History Unit on 24 October 1946, hereafter Rowell Interview; these and other documents on dive bombing are in "Aircraft—Bombers" Subject File, HRS, HMD.

38 MGen Fred S. Robillard, USMC (Ret), As Robie Remembers (Bridgeport, Conn.: Wright Investors' Service, 1969), pp. 45–58.

39 MajGenCmdt, Annual Reports, 1927 and 1929; LtGen William J. Wallace, USMC (Ret), Interview by Oral History Unit, dtd 2 and 13 February 1967 (Oral History Coll, HMD), transcript, pp. 35–40.

40 Quoted in Sherrod, Marine Air History, p. 28.

41 MajGenCmdt, Annual Report, 1927. Rowell Interview.

42 The account of the battle of Ocotal is drawn from the following sources: Maj Ross E. Rowell, USMC, "Annual Report of Aircraft Squadrons, 2nd Brigade, 1 July 1927 to 20 June 1928," MCG, v. XIII, No. 4 (December 1928), hereafter Rowell, "Nicaragua Report, 1928"; Rowell Interview; Headquarters Marine Corps, Division of Operations and Training Report, "Combat Operations in Nicaragua," MCG, v. XIV, No. 1 (March 1929), hereafter Div Ops, "Nicaragua Report, 1927–1928."

43 Rowell Interview.

44 DivOps, "Nicaragua Report, 1927–1928,"p. 20.

45 MajGenCmdt, Annual Report, 1928.

46 Rowell, "Nicaragua Report, 1928," p. 254; Schilt's Flight Log, listing his trips into and out of Quilali is in the Christian F. Schilt Papers (Collections Unit, Marine Corps Museums).

47 "Combat Operations in Nicaragua," MCG, v. XIV, No. 3 (September 1929), pp. 177–179; Rowell Interview.

48 Schilt Interview, p. 72.

CHAPTER IV
Marine Aviation Comes of Age, 1930–1940

1 MajGenCmdt, Annual Report, 1931.

2 MajGenCmdt, Annual Report, 1931, 1932, 1933.

3 Clifford, Progress and Purpose, p. 45.

4 MajGenCmdt, Annual Report, 1934.

5 Clifford, Progress and Purpose, pp. 46–48, 58–59; Peter A. Isely and Philip A. Crowl, The U.S. Marines

and Amphibious War: Its Theory and Its Practice in the Pacific (Princeton, N.J.: 1951), pp. 40–41; Gen Vernon E. Megee, USMC (Ret), ltr to Dir, History & Museums Div, dtd 24 Apr 1975, Comment File, Brief History of Marine Corps Aviation.

6 Quoted in Sherrod, *Marine Air History*, pp. 31–32.

7 Robert Sherrod, "Marine Corps Aviation: The Early Days, Part I," *MCG*, v. XXXVI, No. 6 (May 1952), p. 59.

8 Unless otherwise noted, the information in the rest of this section is drawn from Smith, *Aviation Organization*, pp. 12–16.

9 MajGenCmdt, Annual Report, 1935.

10 MajGenCmdt, Annual Report, 1937, 1938.

11 MajGenCmdt, Annual Report, 1935, 1939, 1940.

12 Except where otherwise noted, all material on the development of the reserve is taken from Smith, *Aviation Reserve*, pp. 42–48.

13 MajGenCmdt, Annual Report, 1935, 1936, 1939.

14 Day Interview, pp. 42–45. See also Torrey Interview, pp. 24–27.

15 MajGenCmdt, Annual Report, 1934, 1935, 1936.

16 BGen Edward C. Dyer, USMC (Ret), Interview by Marine Corps Oral History Unit dtd 7, 9, 19, and 20 August 1968 (Oral History Coll, HMD), transcript, pp. 35–37.

17 *Ibid.*, pp. 31, 36–39, 50–51.

18 *Ibid.*, pp. 35–36, See also pp. 39–44.

19 *Ibid.*, pp. 49, 84–85; MajGenCmdt, Annual Report 1939.

20 Matt and Robinson, *Navy and Marine Fighters*, pp. 47–53, 68–76; Larkins, *USMC Aircraft*, pp. 44, 64, 74, 83, 94, 96. The quotation on the F4B-4 is from Megee Comments.

21 Lynn Montross, "The Marine Autogyro in Nicaragua," *MCG*, v. XXXVII, No. 2 (February 1953), pp. 56–61.

22 MajGenCmdt Annual Report, 1930. For each year's racing and air show activity, See *Ibid.*, 1930–1939.

23 MajGenCmdt, Annual Report 1938 and 1939.

24 All data on amphibious exercises which follows is from Isely and Crowl, *Marines and Amphibious War*, pp. 45–56.

25 Clifford, *Progress and Purpose*, p. 59.

26 Quoted in Isely and Crowl, *Marines and Amphibious War*, pp. 58–59.

APPENDIX A

Directors of Marine Corps Aviation, through Pearl Harbor*

Major Alfred A. Cunningham	17 Nov 1919–12 Dec 1920
Lieutenant Colonel Thomas C. Turner	13 Dec 1920– 2 Mar 1925
Major Edwin H. Brainard	3 Mar 1925– 9 May 1929
Colonel Thomas C. Turner	10 May 1929–28 Oct 1931
Major Roy S. Geiger	6 Nov 1931–29 May 1935
Colonel Ross E. Rowell	30 May 1935–10 Mar 1939
Brigadier General Ralph J. Mitchell	11 Mar 1939–29 Mar 1943

*On 1 April 1936 the title of Officer-In Charge, Aviation, was changed to Director of Aviation.

APPENDIX B

First 100 Marine Corps Aviators

Number	Name	Date of Designation	Naval Aviator Number
1	Alfred Austell Cunningham	17 Sep 1915	5
2	Bernard Lewis Smith	1 Jul 1914	6
3	William Maitland McIlvain	10 Mar 1915	12
4	Francis Thomas Evans	9 Mar 1916	26
5	Roy Stanley Geiger	9 Jun 1917	49
6	David Lukens Shoemaker Brewster	5 Jul 1917	55
7	Edmund Gillette Chamberlain	9 Oct 1917	96 1/2 & 768
8	Russell Alger Presley	9 Nov 1917	100 3/4 & 769
9	Doyle Bradford	5 Nov 1917	111 1/2
10	Clifford Lawrence Webster	5 Nov 1917	112 1/2
11	Arthur Houston Wright	6 Dec 1917	148 & 803
12	Herman Alexander Peterson	2 Nov 1917	163 1/2
13	George McCully Laughlin III	12 Dec 1917	165 & 790
14	Charles Burton Ames	21 Dec 1917	193
15	John Howard Weaver	21 Jan 1918	251 & 794
16	Alvin Lochinvar Prichard	21 Jan 1918	279
17	George Conan Willman	22 Jan 1918	299 & 795
18	Herbert Dalzell Elvidge	12 Mar 1918	424
19	Hazen Curtis Pratt	8 Mar 1918	426
20	Sidney "E" Clark	8 Mar 1918	442 & 800
21	Frederick Commodore Schley	8 Mar 1918	443 & 801
22	Charles Alfred Needham	14 Mar 1918	444
23	John Bartow Bates	25 Mar 1918	449
24	Ralph Talbot	10 Apr 1918	449
25	Thomas Carrington Comstock	26 Mar 1918	473 & 789
26	Francis Osborne Clarkson	28 Mar 1918	474 & 788
27	Guy Mowrey Williamson	25 Mar 1918	477
28	Grover Cleveland Alder	25 Mar 1918	479
29	Edward Kenealy	23 Mar 1918	480
30	Donald Newell Whiting	1 Apr 1918	503
31	Howard Albert Strong	2 Apr 1918	505
32	John Parke McMurran	1 Apr 1918	508 & 791
33	James Kendrick Noble	1 Apr 1918	510 & 792
34	Vincent Case Young	1 Apr 1918	519
35	Province Law Pogue	19 Jun 1918	522 & 782
36	Duncan Hugh Cameron	26 Mar 1918	527 & 787
37	George Fred Donovan	26 Mar 1918	532 & 798
38	William Herbert Derbyshire	28 Feb 1918	533 & 770
39	Frederick Brock Davy	28 Feb 1918	534 & 771
40	Douglas Bennett Roben	14 Mar 1918	535 & 774
41	Arthur Hallett Page, Jr.	14 Mar 1918	536 & 775
42	Gove Compton	14 Mar 1918	537 & 773
43	Thomas James Butler	10 Apr 1918	541 & 786
44	Thomas Rodney Shearer	4 Apr 1918	559
45	Ford Ovid Rogers	14 Apr 1918	560
46	Homer Carter Bennett	11 Apr 1918	562 & 797
47	John Edmond Powell	4 Apr 1918	563
48	William Morrison Barr	8 Apr 1918	567 & 799
49	Harry Eldridge Stovall	11 Apr 1918	568
50	Harvey Byrd Mims	4 Dec 1917	576
51	Winfield Scott Shannon	17 Apr 1918	583
52	Everett Robert Brewer	17 Apr 1918	585

Number	Name	Date of Designation	Naval Aviator Number
53	John George Estill Kipp	17 Apr 1918	586
54	Frederick Louis Kolb	17 Apr 1918	587
55	George Franklin Kremm	17 Apr 1918	588
56	Jesse Arthur Nelson	17 Apr 1918	589
57	Herman Judson Jesse	17 Apr 1918	590
58	William Webster Head	17 Apr 1918	591
59	Gustav Henry Kaemmerling	17 Apr 1918	592
60	Jesse Floyd Dunlap	17 Apr 1918	593
61	Trevor George Williams	17 Apr 1918	594
62	Clyde Noble Bates	17 Apr 1918	595
63	Melville Edward Ingalls Sullivan	17 Apr 1918	596
64	Francis Patrick Mulcahy	17 Apr 1918	597
65	Benjamin Louis Harper	17 Apr 1918	598
66	Walter Harold Batts	17 Apr 1918	599
67	Henry Teasdale Young	17 Apr 1918	600
68	Karl Schmolsmire Day	17 Apr 1918	601
69	Fred Sevier Robillard	17 Apr 1918	602
70	Melchior Borner Trelfall	17 Apr 1918	603
71	Harold Cornell Major	17 Apr 1918	604
72	Robert Sidney Lytle	17 Apr 1918	605
73	Thomas Caldwell Turner	14 Mar 1918	772
74	Kenneth Brown Collings	26 Mar 1918	776
75	Donald Buford Cowles	4 Apr 1918	777
76	Maco Stewart, Jr.	4 Apr 1918	778
77	Henry Sidney Ehret, Jr.	6 Apr 1918	779
78	Raymond Joseph Kirwan	8 Apr 1918	780
79	Frank Nelms, Jr.	19 Jun 1918	781
80	Harvey Chester Norman	23 May 1918	783
81	Delmar Leighton	23 May 1918	784
82	John Thomas Brecton	11 Apr 1918	785
83	William Wheelwright Torrey	22 Mar 1918	793
84	Joseph White Austin	23 Mar 1918	796
85	Bunn Gradon Barnwell	28 May 1918	804
86	Walter Josephs Willoughby	19 Jun 1918	805
87	Chester Julius Peters	19 Jun 1918	806
88	Roswell Emory Davis	19 Jun 1918	807
89	Horace Wilbur Leeper	25 Jun 1918	808
90	Byron Brazil Freeland	25 Jun 1918	809
91	Robert James Paisley	19 Jun 1918	810
92	Charles Thomas Holloway II	1 Jul 1918	811
93	Frank Henry Fleer	2 Jul 1918	812
94	Maurice Kingsley Heartfield	2 Jul 1918	813
95	Robert James Archibald	8 Jul 1918	814
96	Arthur Judson Sherman	8 Jul 1918	815
97	Philip William Blood	8 Jul 1918	816
98	Albert Aloysius Kuhlen	28 Jun 1918	817
99	Earl Francis War	30 Jun 1918	818
100	August Koerbling	1 Jul 1918	819

NOTE: Aviators with two designation numbers generally transferred from the Navy to the Marine Corps, receiving a second number from the Marines. The lower number is used to establish precedence. Numbers with fractions resulted from several aviators being given the same designation number. Also, dates of designation should not be confused with dates of precedence, which are reflected by naval aviator numbers and are often much earlier than designation dates.

APPENDIX C

Marine Corps Aircraft, 1913-1940

Designation	Type	Year Assigned	Manufacturer and Name	Engine Type and Horsepower	Dimensions Length and Span	
1. AX-1	Bat Boat, 1-engine 2-crew, biplane	1913	Curtiss	Curtiss 90 h.p.	27'2"	37'1"
2. JN-4B	Trainer, 1-engine, 2-crew, biplane	1917	Curtiss "Jenny"	Curtiss DXX 100 h.p.	27'4"	43'3"
3. H-12	Patrol, 2-engine, 2-crew, biplane, flying boat	1918	Curtiss	2 Liberty 42 cyl. 300 h.p.	46'1"	95'
4. H-16	Patrol, 2-engine, 2-crew, 4-place, biplane, flying boat	1918	Curtiss, Naval Aircraft Factory; and others.	2 Liberty 12 cyl. 300 h.p.	46'1"	95'
5. HS-2	Patrol, 1-engine, 2-crew, biplane, flying boat	1918	Curtiss; Standard; Naval Aircraft Factory; Lowe, Willard, and Fowler; and others.	Liberty 12 cyl. 330 h.p.	39'	74'
6. HS-2L	Patrol, 1-engine, 2-crew, biplane	1918	Curtiss; Lowe, Willard, and Fowler; and others.	Liberty 12 cyl. 360 h.p.	39'	74'
7. Kirkham Fighter	Experimental fighter, 1-engine, 2-place, triplane	1918	Curtiss	Kirkham 400 h.p.	23'	31'10"
8. N-9	Trainer, 1-engine, 2-place, 1 float, biplane, seaplane	1918	Curtiss; Burgess	Curtiss	30'10"	53'4"
9. R-6	Trainer, 1-engine, 2-place, biplane, seaplane.	1918	Curtiss	Curtiss V-2 200 h.p.	33'5"	57'1"
10. DH-4	Observation, 1-engine, 2-crew, biplane	1920	Dayton-Wright	Liberty 12 cyl. 360 h.p.	30'2"	42'6"
11. DH-9A	Observation bomber, 1-engine, 2-crew, biplane	1918	British Aircraft Manufacturing Co.	Liberty 12 cyl. 400 h.p.	30'3"	45'11"

This list is reproduced with amendments from Historical Branch, G-3 Division, HQMC, *Marine Corps Aircraft 1913–1965* (Washington, DC: HMD, 1967, rev. ed.). The amendments include the addition of specifications for the DH-9A taken from W. M. Lamberton, comp., and E. F. Cheeseman, ed., *Reconnaissance and Bomber Aircraft of the 1914–1918 War* (Los Angeles: Aero Publishers Inc., 1962).

Designation	Type	Year Assigned	Manufacturer and Name	Engine Type and Horsepower	Dimensions Length and Span	
12. DH-4B	Observation, 1-engine, 2-crew, biplane	1920	U.S. Army	Liberty 42 cyl. 400 h.p.	30'2"	42'6"
13. E-1 "M" Defense	Fighter, 1-engine, biplane	1920	Standard	LeRhone 80 h.p.	18'11"	24'
14. HS-1	Patrol, 1-engine, 2-crew, biplane	1920	Curtiss	Liberty 12 cyl. 360 h.p.	38'6"	62'1"
15. JN-4	Trainer, 1-engine, 2-crew, biplane	1920	Curtiss "Jenny"	Curtiss OXX 100 h.p.	27'1"	43'7"
16. JN-6-HG-1	Trainer, 1-engine, 2-crew, biplane	1920	Curtiss "Jenny"	Hispano 150 h.p.	27'	43'3"
17. VE-7	Trainer, 1-engine, 2-crew, biplane	1920	Lewis and Vought	Hispano E-2 180 h.p.	24'5"	34'1"
18. Fokker C-1	Fighter, 1-engine, 2-crew, biplane	1921	Netherlands Aircraft Company	B.M.W. 243 h.p.	23'8"	34'10"
19. Fokker D-7	Fighter, 1-engine, 1-crew, biplane	1921	Fokker	Packard 350 h.p.	23'	27'6"
20. VE-7G	Trainer, 1-engine, 2-crew, biplane, seaplane	1921	Naval Aircraft Factory	Hispano E-2 480 h.p.	24'5"	34'1"
21. VE-7SF	Fighter trainer, 1-engine, 1-crew, land, biplane	1921	Vought	Hispano E-2 480 h.p.	24'5"	34'1"
22. DH-4B-1	Observation, 1-engine, 2-crew, biplane	1922	U.S. Army	Liberty 12 cyl. 400 h.p.	30'2"	42'6"
23. F-5-L	Patrol bomber, scout, 2-engine, 2-crew, 5-place, biplane, flying boat	1922	Naval Aircraft Factory; Curtiss; and others	2 Liberty 12 cyl. 360 h.p.	49'4"	103'9"
24. MB-3	Fighter, 1-engine, 1-crew, biplane	1922	Thomas-Morse	Hispano 300 h.p.	20'	26'
25. MBT	Torpedo bomber, 2-engine, 3-crew, biplane	1922	Martin	2 Liberty 12 cyl. 400 h.p.	46'4"	71'5"
26. MT	Torpedo bomber, 2-engine, 3-crew, biplane	1922	Martin	2 Liberty 12 cyl. 400 h.p.	46'4"	71'5"
27. DT-2	Torpedo bomber, 1-engine, 2-crew, convertible (land or sea), biplane	1923	Douglas; Naval Aircraft Factory; Lowe, Willard, and Fowler	Liberty 12 cyl. 450 h.p.	37'8"	50'
28. JN-4H	Trainer, 1-engine, 2-crew, biplane	1923	Curtiss "Jenny"	Hispano Suiza 150 h.p.	27'	43'8"
29. T3M-1	Torpedo bomber, 1-engine, 3-crew, 2-float, convertible, lower wing had wider span	1923	Martin	Wright 575 h.p.	42'9"	56'7"
30. VE-9	Observation, 1-engine, 2-crew, biplane	1923	Vought	Wright E-3 180 h.p.	24'6"	34'1"

Designation	Type	Year Assigned	Manufacturer and Name	Engine Type and Horsepower	Dimensions Length and Span	
31. DH-4B-2	Observation, 1-engine, 2-crew, biplane	1925	Naval Aircraft Factory	Liberty 12 cyl. 400 h.p.	30'2"	42'5"
32. JN-6H	Trainer, 1-engine, 2-crew, biplane	1925	Curtiss "Jenny"	Hispano 180 h.p.	26'11"	43'7"
33. JN-6H-B	Same configuration as number 31.					
34. O2B-1	Observation, 1-engine, 2-crew, biplane	1925	Boeing	Liberty 400 h.p.	30'2"	42'6"
35. TW-3	Trainer, 1-engine, 2-crew, 1-float, biplane, convertible	1925	Dayton Wright "Chummy"	Wright 180 h.p.	25'11"	34'10"
36. VE-7H	Trainer, 1-engine, 2-crew, 1-float, biplane, seaplane	1925	Vought	Wright E-2 180 h.p.	24'5"	34'2"
37. F6C-3	Fighter, 1-engine, 1-crew, 2-float, biplane, convertible	1926	Curtiss "Hawk"	Curtiss D-12 400 h.p.	22'8"	31'6"
38. FB-1	Fighter, 1-engine, 1-crew, biplane	1926	Boeing	Curtiss D-12 400 h.p.	23'6"	32'
39. NB-1	Trainer, 1-engine, 2-crew, 1-float, biplane, convertible	1926	Boeing	Wright J-4 200 h.p.	28'9"	36'10"
40. NB-2	Trainer, 1-engine, 2-crew, 1-float, biplane, convertible	1926	Boeing	Wright E-4 180 h.p.	28'9"	36'10"
41. NY-1	Trainer, 1-engine, 2-crew, 1-float	1926	Consolidated	Wright J-5 200 h.p.	31'5"	34'6"
42. OD-1	Observation, 1-engine, 2-crew, biplane	1926	Douglas	Packard 4A-1500 500 h.p.	28'8"	39'8"
43. OL-2	Observation, 1-engine, 2-crew, biplane	1926	Loening	Liberty 400 h.p.	33'10"	45'
44. XS-1	Scout, 1-engine, 1-crew, 2-float, biplane, seaplane	1926	Cox-Klemin	Kinner, 5 RA 84 h.p.	18'2"	18'
45. F6C-1	Fighter, 1-engine, 2-crew, 2 float, biplane, convertible	1927	Curtiss "Hawk"	Curtiss D-12 400 h.p.	22'8"	31'6"
46. F6C-4	Fighter, 1-engine, 2-crew, 2-float, biplane, convertible	1927	Curtiss "Hawk"	Pratt & Whitney R-1340 410 h.p.	22'5"	31'6"
47. O2Y-1	Observation, 1-engine, 2-crew, 1-float, biplane, convertible	1927	Vought "Corsair"	Pratt & Whitney R-1300 425 h.p.	24'8"	34'6"
48. OL-4	Observation, 1-engine, 3-crew, biplane, amphibian	1927	Loening	Liberty 400 h.p.	35'1"	45'
49. OL-6	Observation, 1-engine, 3-crew, biplane, amphibian	1927	Loening	Packard 2A-1500 475 h.p.	35'4"	45'
50. TA-1	Transport, 3-engine, 2-crew, high wing monoplane	1927	Atlantic; Fokker	3 Wright J-5 220 h.p.	49'1"	63'4"

Designation	Type	Year Assigned	Manufacturer and Name	Engine Type and Horsepower	Dimensions Length and Span	
51. XF6C-5	Experimental fighter, 1-engine, 1-crew, 2-float, biplane, convertible	1927	Curtiss "Hawk"	Pratt & Whitney R–1700 525 h.p.	25'5"	31'6"
52. F7C-1	Fighter, 1-engine, 1-crew, 1-float, convertible	1928	Curtiss "Sea Hawk"	Pratt & Whitney, R–1340–B 450 h.p.	22'2"	32'8"
53. F8C-1	Fighter, 1-engine, 2-crew, biplane	1928	Curtiss "Helldiver"	Pratt & Whitney R–1340–B 450 h.p.	25'11"	32'
54. F8C-3	Same configuration as number 52.					
55. NY-1B	Trainer, 1-engine, 2-crew, 1-float, hiplane, convertible	1928	Consolidated	Wright J–5 220 h.p.	31'4"	34'6"
56. O2B-2	Observation, 1-engine, 2-crew, biplane	1928	Naval Aircraft Factory	Liberty 400 h.p.	30'1"	42'5"
57. OC-1	Observation, 1-engine, 2-crew, biplane	1928	Curtiss "Falcon" (redesignated from F8C-1)	Pratt & Whitney R–1340 410 h.p.	28'	38'
58. OC-2	Same configuration as number 56.					
59. OL-8	Observation, 1-engine, 2-crew, biplane, amphibian	1928	Loening	Pratt & Whitney R–1300 425 h.p.	34'9"	45'
60. TA-2	Transport, 3-engine, 2-crew, monoplane	1928	Atlantic; Fokker	2 Wright R–790A 300 h.p. 1 Pratt & Whitney 450 h.p.	48'7"	72'10"
61. UO-1	Observation, 1-engine, 2-crew, 1-float, biplane convertible	1928	Vought "Corsair"	U–8–D 250 h.p.	29'3"	34'1"
62. UO-5	Observation, 1-engine, 2-crew, biplane, convertible	1928	Vought "Corsair"	Wright J–5 220 h.p.	28'4"	34'4"
63. XOL-8	Experimental observation, 1-engine, 3-crew, biplane, amphibian	1928	Loening	Pratt & Whitney R–1300 425 h.p.	34'9"	45'
64. FB-5	Fighter, 1-engine, 1-crew, biplane	1929	Boeing	Packard 12A–1500 475 h.p.	23'2"	32"
65. JR-2	Transport, 3-engine, 2-crew, 10-passenger, high wing monoplane	1929	Ford "Tin Goose"	3 Wright R–790A 300 h.p.	49'10"	74"
66. OL-3	Observation, 1-engine, 3-crew, biplane, amphibian	1929	Loening	Packard 2A-2500 475 h.p.	35'1"	45'
67. XHL-1	Experimental transport 1-engine, 2-crew, biplane, amphibian, cabin-ambulance	1929	Loening	Pratt & Whitney R–1690 525 h.p.	34'9"	46'10"

Designation	Type	Year Assigned	Manufacturer and Name	Engine Type and Horsepower	Dimensions Length and Span	
68. F8C-5	Fighter, 1-engine, 2-crew, biplane	1930	Curtiss "Helldiver"	Pratt & Whitney R–1340C 450 h.p.	25'11"	32'
69.	Same configuration as number 64.					
70. O2U-4	Observation, 1-engine, 2-crew, 1-float, bi-plane, convertible	1930	Vought "Corsair"	Pratt & Whitney R–1340C 450 h.p.	30'	36'
71. TA-3	Transport, 3-engine, 2-crew, high wing monoplane	1930	Atlantic; Naval Air-craft Factory	3 Wright R–975 300 h.p.	48'1"	63'4"
72. XN2B-1	Experimental trainer, 2-crew, biplane	1930	Boeing	Wright R–540 165 h.p.	25'8"	35'
73. XOC-3	Experimental observation, 1-engine, 2-crew, biplane	1930	Curtiss "Falcon"	Pratt & Whitney R–1340C 450 h.p.	28'	38'
74. NT-1	Trainer, 1-engine, 2-crew, biplane	1931	New Standard	Kinner K–5 415 h.p.	24'7"	30'
75. O2C-1	Same configuration as number 67.					
76. O3U-2	Observation, 1-engine, 2-crew	1931	Vought "Corsair"	Pratt & Whitney R–1690C 600 h.p.	26'	36'
77. OL-9	Observation, 1-engine, 2-crew, biplane, amphibian	1931	Loening	Pratt & Whitney R–1340C 450 h.p.	34'9"	45'
78. OP-1	Observation, 1-engine, 2-crew, autogiro	1931	Pitcairn	Wright R–975 300 h.p.	23'1"	Rotor 30'3"
79. RA-3	Same configuration as number 70.					
80. RC-1	Transport, 2-engine, 2-crew, ambulance, high-wing, boxtail, monoplane	1931	Curtiss-Wright "Kingbird"	2 Wright R–975 300 h.p.	34'10"	54'6"
81. RR-2	Same configuration as number 64.					
82. RR-3	Transport, 3-engine, 2-crew, 10-passenger, high-winged monoplane	1931	Ford "Tin Goose"	3 Pratt & Whitney R–1340-C 450 h.p.	50'3"	77'10"
83. RS-1	Transport, 2-engine, 2-crew, 7-passenger, high-wing, parasol wing, amphibian	1931	Sikorsky	2 Pratt & Whitney R–1860 575 h.p.	45'2"	79'9"
84. RS-3	Transport, 2-engine, 2-crew, 8-passenger, biplane, amphibian	1931	Sikorsky	2 Pratt & Whitney R–1340-C 575 h.p.	40'3"	71'8"

Designation	Type	Year Assigned	Manufacturer and Name	Engine Type and Horsepower	Dimensions Length and Span	
85. T4M-1	Torpedo bomber, 1-engine, 3-crew, biplane, convertible	1931	Martin	Pratt & Whitney R-1690 525 h.p.	37'8"	53'
86. F4B-4	Fighter, 1-engine, 1-crew, biplane, land-carrier	1932	Boeing	Pratt & Whitney R–1340–D 500 h.p.	20'4"	30'
87. RR-5	Same configuration as number 81.					
88. SU-2	Scout, 1-engine, 2-crew, biplane, land-carrier	1932	Vought "Corsair" (formerly O3U-4)	Pratt & Whitney R–1690–C 600 h.p.	26'	36'
89. F3B-1	Fighter, 1-engine, 1-crew, 1-float, biplane, convertible, land-battleship-carrier	1933	Boeing	Pratt & Whitney R–1340–B 450 h.p.	24'10"	33"
90. N2C-2	Trainer, 1-engine, 2-crew, biplane, convertible	1933	Curtiss "Fledgling"	Wright R–760A 240 h.p.	27'9"	39'1"
91. RE-3	Transport, 1-engine, 2-crew, 4-passenger, high-wing, monoplane	1933	Bellanca "Pacemaker"	Pratt & Whitney R–1340–CD 450 h.p.	27'10"	46'4"
92. SU-3	Scout, 1-engine, 2-crew, biplane, land-carrier	1933	Vought "Corsair"	Pratt & Whitney R–1690C, 600 h.p.	26'	36"
93. F4B-3	Fighter, 1-engine, 1-crew, biplane, land-carrier	1934	Boeing	Pratt & Whitney R–1340D 500 h.p.	20'	30'
94. JF-1	Utility, 1-engine, 2-crew, biplane, amphibian boat hull	1934	Grumman "Duck"	Pratt & Whitney R–1830–62	14'4"	39'
95. R2D-1	Transport, 2-engine, 2-crew, 14-passenger, low-wing, monoplane	1934	Douglas	2 Wright R–1820–12 725 h.p.	62'	85'
96. R4C-1	Transport, 2-engine, 2-crew, 14-passenger, biplane	1934	Curtiss-Wright "Condor"	2 Wright R–1820–12 725 h.p.	50'3"	82'
97. RR-4	Transport, 3-engine, 2-crew, 10-passenger, all metal cabin, high-wing monoplane.	1934	Ford "Tin Goose"	3 Pratt & Whitney R–1340–96	50'3"	77'10"
98. SU-1	Scout, 1-engine, 2-crew, biplane, land-carrier	1934	Vought "Corsair" (redesignated from O3U-2)	Pratt & Whitney R–1690–40 600 h.p.	26'3"	36'
99. BG-1	Bomber, 1-engine, 2-crew, biplane, staggered wing, land-carrier	1935	Great Lakes	Pratt & Whitney R–1535–66 700 h.p.	28'9"	36'
100. JF-2	Utility, 1-engine, 2-crew, biplane, amphibian, boat hull	1935	Grumman "Duck"	Pratt & Whitney R–1820–62 700 h.p.	14'4"	39'

Designation	Type	Year Assigned	Manufacturer and Name	Engine Type and Horsepower	Dimensions Length and Span	
101. O3U-6	Observation scout, 1-engine, 2-crew, biplane, convertible, land or sea	1935	Vought "Corsair"	Pratt & Whitney R–1340–12 550 h.p.	27'2"	36'
102. RD-3	Transport, 2-engine, 2-crew, 7-passenger, high-wing, monoplane, amphibian, boat hull	1935	Douglas	2 Pratt & Whitney R–1340–96 450 h.p.	45'2"	60'
103. SOC-1	Scout observation, 1-engine, 2-crew, biplane, convertible equipped for catapult	1935	Curtiss	Pratt & Whitney R–1340–18 550 h.p.	26'10"	36'
104. F2A-1	Fighter, 1-engine, 1-crew, mid-wing monoplane	1936	Brewster "Buffalo"	Pratt & Whitney 850 h.p.	26'	35'
105. F3F-1	Fighter, 1-engine, 1-crew, biplane, land-carrier	1936	Grumman	Pratt & Whitney R–1535–84 650 h.p.	23'5"	32'
106. O3U-1	Observation, 1-engine, 2-crew, biplane, convertible, battleship-carrier	1936	Vought "Corsair"	Pratt & Whitney R–1340–96 450 h.p.	29'11"	36'
107. RD-2	Transport, 2-engine, 2-crew, 7-passenger, high-wing monoplane, boat hull	1936	Douglas	2 Pratt & Whitney R–1340–96 450 h.p.	45'3"	60'
108. SU-4	Scout, 1-engine, 2-crew, biplane, land-carrier	1936	Vought "Corsair"	Pratt & Whitney R–1690–42 600 h.p.	28'	36'
109. XBG-1	Experimental bomber, 1-engine, 2-crew, biplane, carrier	1936	Great Lakes	Pratt & Whitney R–1535–66 700 h.p.	33'9"	36'
110. F2F-1	Fighter, 1-engine, 1-crew, biplane, land-carrier	1937	Grumman	Pratt & Whitney R–1535–72 750 h.p.	21'2"	28'6"
111. F3F-2	Fighter, 1-engine, 1-crew, biplane	1937	Grumman	Pratt & Whitney R–1535–84 650 h.p.	23'2"	32'
112. J2F-1	Utility, 1-engine, 2-crew, biplane, amphibian, boat hull	1937	Grumman "Duck"	Pratt & Whitney R–1820–08 750 h.p.	34'	39'
113. JO-2	Transport, 2-engine, 2-crew, 6-passenger, low-wing, monoplane	1937	Lockheed	2 Pratt & Whitney Aircraft R–985–48 400 h.p.	36'4"	49'6"
114. SBC-3	Scout bomber, 1-engine, 2-crew, biplane, carrier-land	1937	Curtiss "Helldiver"	Wright Whitney Aircraft R–1535–94 750 h.p.	28'	34'

Designation	Type	Year Assigned	Manufacturer and Name	Engine Type and Horsepower	Dimensions Length and Span	
115. XB2G-1	Experimental bomber, 4-engine, 2-crew, biplane, land-carrier	1937	Great Lakes	Pratt & Whitney Aircraft, R–1535–94 750 h.p.	28'10"	36'
116. XF13C-3	Experimental fighter, 1-engine, 1-crew, high-wing, monoplane, land-carrier	1937	Curtiss	Wright XR–1510–12 700 h.p.	26'4"	35'
117. J2F-2	Utility, 1-engine, 2-crew, biplane, amphibian, boat hull	1938	Grumman "Duck"	Wright R–1820–30 750 h.p.	33'	39'
118. JRS-1	Utility transport, 2-engine, 5-crew, parasol wing, high-wing, monoplane, flying boat	1938	Sikorsky	Wright R–1690–52 600 h.p.	51'1"	86'
119. O3U-3	Observation, 1-engine, 2-crew, biplane, convertible	1938	Vought "Corsair"	Pratt & Whitney Corp. R–1340–12 600 h.p.	31'	36'
120. SB2U-1	Scout bomber, 1-engine, 2-crew, low-wing, monoplane	1938	Vought-Sikorsky "Vindicator"	Pratt & Whitney Twin-Wasp 750 h.p.	34'	42'
121. SOC-3	Scout observation, 4-engine, 2-crew, biplane, convertible, catapult	1938	Curtiss Wright "Seagull"	Pratt & Whitney R–1340–22 550 h.p.	31'1"	36'
122. TG-1	Torpedo bomber, 4-engine, 3-crew, 2-float, biplane, convertible, carrier	1938	Great Lakes	Pratt & Whitney R–1690–28 525 h.p.	34'8"	53'
123. F3F-3	Fighter, 1-engine, 1-crew, biplane, land-carrier	1939	Grumman	Wright Cyclone 750 h.p.	23'3"	32'
124. J2F-2A	Utility, 1-engine, 4-crew, biplane, amphibian	1939	Grumman "Duck"	Wright R–1820–F5A 775 h.p.	34'	39'
125. J2F-4	Utility, 1-engine, biplane, amphibian, boat hull	1939	Grumman "Duck"	Wright Cyclone 725 h.p.	34'	39'
126. JRF-1A	Utility, 2-engine, 4-crew, high-winged, monoplane, boat hull	1939	Grumman "Goose"	2 Pratt & Whitney Wasp Junior SB Radial 450 h.p.	38'4"	49'
127. SBC-4	Scout bomber, 1-engine, 2-crew, biplane, land-carrier	1939	Curtiss "Helldiver"	Wright Cyclone R–1820–C–3 875 h.p.	27'5"	34'
128. XSBC-4	Experimental scout-bomber, 1-engine, 2-crew, biplane, land-carrier	1939	Curtiss "Helldiver"	Wright Cyclone R–1820–C–3 875 h.p.	275'5"	34'
129. R3D-2	Transport, 2-engine, 4-crew, high-wing monoplane	1940	Douglas	2 Wright Cyclones GR–1820–G102A 1100 h.p.	62'2"	78'

Designation	Type	Year Assigned	Manufacturer and Name	Engine Type and Horsepower	Dimensions Length and Span	
130. SBD-1	Scout bomber, 1-engine, 2-crew, low-wing, monoplane	1940	Douglas "Dauntless"	Wright Cyclone R–1820 950 h.p.	32'	41'
131. SNJ-2	Scout trainer, 1-engine, 2-crew low-wing, monoplane	1940	North American "Texan"	Wright Whirlwind 400 h.p.	28'11"	42'
132. F2A-3	Fighter, 1-engine, 1-crew, mid-wing, monoplane, carrier	1940	Brewster "Buffalo"	Pratt & Whitney F–1820–40 1000 h.p.	26'4"	35'

NOTE: In its earliest years, Marine aviation had no system of aircraft type and manufacturer identification. For example, the HS-2L was built by Curtiss; Lowe, Willard, and Fowler; and others. In 1922, a system was devised whereby the first letter indicated manufacturer, the second letter the plane's mission, and an appended number for modifications. A number between the letters stood for the order or model number of the designer's aircraft in the same class—the first design "1" was omitted. Thus a U2O–1 indicates a (U) Vought, (2) second design of, (O) observation aircraft, with (1) its first modification. In 1923 the system was reversed so that the mission letter came first and the manufacturer's letter came second. This system remained in effect through the period covered in this history.

TYPE LETTERS

A—Attack; ambulance
B—Bomber
F—Fighter
C—Transport (single engine)
H—Helicopter; hospital
J—Transport and general utility
JR—Utility-transport
N—Trainer
O—Observation
OS—Observation-scout

P—Patrol
PB—Patrol bomber
R—Transport (Multiengine)
S—Scout
SB—Scout bomber
SN—Scout trainer
SO—Scout observation
T—Torpedo bomber; trainer
TB—Torpedo bomber
U—Utility
X—Experimental

MANUFACTURERS' SYMBOLS

Date indicates first year that particular manufacturer's symbol appeared in the designation of an aircraft assigned to the Marines.

A—Atlantic (1927)
A—Brewster (1936)
B—Beech (1941)
B—Boeing (1925)
C—Curtiss (Curtiss-Wright) (1926)
D—Douglas (1923)
E—Bellanca (1923)
F—Grumman (1934)
G—Great Lakes (1935)
J—North American (1940)
L—Loening (1926)

M—Glenn L. Martin (1922)
O—Lockheed (1939)
P—Pitcairn (1931)
P—Spartan (1937)
R—Ford (1929)
S—Sikorsky (1931)
T—New Standard (1931)
U—Vought (1927)
W—Dayton-Wright (1925)
X—Cox-Klemin (1926)
Y—Consolidated (1926)

APPENDIX D

Awards to Marine Officers and Enlisted Men for Aviation Duty, 1912-1940

Medal of Honor

Christian F. Schilt	1stLt	Nicaragua
Ralph J. Talbot	2dLt	World War I
Robert G. Robinson	GySgt	World War I

Navy Cross

Alfred A. Cunningham	Maj	World War I
Roy S. Geiger	Maj	World War I
William M. McIlvain	Maj	World War I
Douglas B. Roben	Maj	World War I
Robert E. Williams	Capt	World War I
Karl S. Day	Capt	World War I
Donald M. Whiting	1stLt	World War I
John R. Whiteside	1stLt	World War I
Arthur H. Wright	1stLt	World War I
Ford O. Rogers	1stLt	World War I
Herman A. Peterson	1stLt	World War I
Eynar F. Olsen	1stLt	World War I
George McC. Laughlin, III	1stLt	World War I
Albert E. Humphreys	1stLt	World War I
Everett R. Brewer	1stLt	World War I
Clyde M. Bates	1stLt	World War I
Fred S. Robillard	1stLt	World War I
Chapin C. Barr	2dLt	World War I
John H. Weaver	2dLt	World War I
Caleb W. Taylor	2dLt	World War I
Harvey C. Norman	2dLt	World War I
Harold A. Jones	2dLt	World War I
John K. McGraw	1st Sgt	World War I
Harry B. Wershiner	GySgt	World War I
Thomas L. McCullough	Sgt	World War I

Distinguished Service Medal

Ross E. Rowell	LtCol	Nicaragua
Francis P. Mulcahy	Capt	World War I
Robert S. Lytle	Capt	World War I
Frank Nelms	2dLt	World War I
Amil Wiman	GySgt	World War I

Distinguished Flying Cross

Thomas C. Turner	Col	Pioneer Flight, 22 Apr 1921
Ross E. Rowell	LtCol	Nicaragua
Ralph J. Mitchell	Maj	Nicaragua
Louis M. Bourne	Maj	Nicaragua
Arthur H. Page	Capt	Pioneer Flight, 2 Jul 1930
Byron F. Johnson	Capt	Nicaragua
Alton N. Parker	Capt	Antarctic
Hayne D. Boyden	1stLt	Nicaragua
Lawson H. M. Sanderson	1stLt	Pioneer Flight, 22 Apr 1921
Basil Bradley	1stLt	Pioneer Flight, 22 Apr 1921
Herbert P. Becker	1stLt	Nicaragua
Frank H. Lamson-Scribner	1stLt	Nicaragua
Frank D. Weir	1stLt	Nicaragua
Charles L. Fike	1stLt	Nicaragua
John N. Hart	1stLt	Nicaragua
John S. E. Young	1stLt	Nicaragua
Michael Wodarczyk	MG	Nicaragua
Albert S. Munsch	MSgt	Nicaragua
Charles W. Rucker	GySgt	Pioneer Flight, 22 Apr 1921
Gordon W. Heritage	SSgt	Nicaragua
Hilmar N. Torner	Sgt	Test Flight, 22 Mar 1932

INDEX

www.ingramcontent.com/pod-product-compliance
Lightning Source LLC
Chambersburg PA
CBHW050354100426
42739CB00015BB/3397